Analog Devices and Circuits 2

Analog Devices and Circuits 2

Analog Circuits

Christian Gontrand

iSTE

WILEY

First published 2024 in Great Britain and the United States by ISTE Ltd and John Wiley & Sons, Inc.

ISTE Ltd
27-37 St George's Road
London SW19 4EU
UK

www.iste.co.uk

John Wiley & Sons, Inc.
111 River Street
Hoboken, NJ 07030
USA

www.wiley.com

Library of Congress Control Number: 2023947046

British Library Cataloguing-in-Publication Data
A CIP record for this book is available from the British Library
ISBN 978-1-78630-900-6

Contents

Preface

At the end of the Second World War, a new technological trend was born: integrated electronics. This relied on the enormous rise of integrable electronic circuits. This field has invaded our societies, and it is far from over!

Since the 1970s, modeling has contributed a great deal to the understanding of basic electric functions, making it possible to carry out "*numerical* experiments", accelerating the effective development of circuits. In these studies, we hope to detect assembly errors as quickly as possible; for example, the arrows of currents, indicated in the diagrams of devices and circuits, allow us to verify that there is no break in the current flow between the plus and minus "rails", typically correcting any "assembly" error of the circuit.

But, of course, this simulation has allowed for significant productivity gains, avoiding, in particular, the effective and overly-upstream manufacture of demonstrators.

This book is in fact a second volume, wherein the first discusses analog devices constituting these circuits. Of course, in this volume we will focus a little more on the "flagship chip": the operational amplifier.

Acknowledgments

The author would like to thank the Union for the Mediterranean (UfM), Euromed University of Fes (UEMF), and Institut National des Sciences Appliquées (INSA) for supporting this study.

November 2023

1

On Analog Circuits

1.1. Introduction: miscellaneous

1.1.1. *SPICE*

SPICE (Simulation Program with Integrated Circuit Emphasis) is the standard for simulating analog circuits. Anyone who does not know this does not belong to the fraternity of electronic circuit analogists.

SPICE was created at the University of California (Berkeley) in the early 1970s by Ron Rohrer's team, including Larry Nagel (see his famous thesis). It later became the standard for analog simulators. Three versions followed one another, including SPICE3, dated 1985.

1.1.1.1. *A brief history of SPICE*

The need for a circuit simulation program, "smart" people with a vision and hardworking teams of students and professionals have all contributed to the realization and evolution of SPICE. A brief history of this powerful simulator is explained below, which is organized primarily according to the different versions of SPICE.

CANCER

– In the early 1970s, Ron Rohrer hoped to develop a simulation program for his work on optimization at the University of California, Berkeley.

– Rohrer's students, including Larry Nagel, created CANCER (Computer Analysis of Nonlinear Circuits Excluding Radiation).

– It performs DC, AC, and transient analysis.

– The devices include diodes (Shockely equations) and bipolar junction transistors (Ebers–Moll equations).

Other simulation programs at the time included ECAP (Electronic Circuit Analysis Program) and IBM's Autonetics TRAC.

SPICE1

– In 1972, Nagel and Pederson launched SPICE1 (Simulation Program with IC Emphasis) in the public domain.

– SPICE became the industry standard simulation tool.

– Bipolar junction transistor models were replaced by Gummel–Poon equations.

– JFET and MOSFET templates were added.

– It was based on nodal analysis.

– It was written in FORTRAN code and runs on large computers.

SPICE2

– Nagel's 1975 version offered significant improvements.

– Modified nodal analysis (MNA), replacing the old analysis, supported voltage sources and inductors from this point onwards.

– Memory was dynamically allocated to accommodate the increasing size and complexity of circuits.

– It has adjustable simulation of time step control speeds.

– The MOSFET and bipolar models were revised and extended.

– SPICE2G.6 (1983) is the latest version of FORTRAN.

At present, it is still available in Berkeley. Many commercial simulators today are based on SPICE2G.6.

SPICE3

– SPICE code was rewritten in programming language C (1985).

– It has a graphical interface to display the results.

– It included polynomial capacitors, inductors and voltage-controlled sources.

– The new version eliminated many convergence problems.

– Models added were as follows: MESFET, lossy transmission line and nonideal switch.

– Improved semiconductor models adapted to smaller transistor geometries.

– It is not backward compatible with SPICE2.

1980s and beyond

– Published commercial versions include HSPICE, IS_SPICE and MICROCAP.

– MicroSim launched PSPICE, the first PC version of SPICE.

– SPICE attracted many more users in industry and academia.

– EDLO is dedicated to RF.

– HICUM is dedicated to microwaves.

– Companies integrated SPICE versions into their "schematics" entry and layout packages (geometry/pattern).

1.1.1.2. *A SPICE program*

SPICE is therefore the essential software for studying analog circuits.

1.1.1.3. *Program example*

The program example is given as follows.

```
I-V characteristics for SS model of CMOS devices
****************************************************************
MODEL NSS NMOS LEVEL =3 RSH=0 TOX=275E-10 LD=0.1E-6 XJ=0.14E-6
+ CJ=1.6E-4 CJSW=1 8E-10 UO550 VTO=1.022 CGSO=1.3E10
+ CGDO=1 3E-10 NSUB=4E15 NFS=1E10
+ VMAX=12E4 PB=0.7 MJ=0.5 MJSW=0.3 THETA=0.06 KAPPA=0.4 ETA=0.14
.MODEL PSS PMOS LEVEL=3RSH=0 TOX=275E-10 LD=0.3E-6 XJ=0.42E-6
+ CJ=7. 7E-4 CJSW=5.4E-10 UO=180 VTO=-1.046 CGSO=4E-10
+ CGDO=4E-10 TPG=-1 NSUB=7E15 NFS=1E10
+ VMAX=12E4 PB=0 . 7 MJ=0 . 5 MJSW=0 . 3 ETA=0 . 06 THETA=0 . 03
KAPPA=0.4
****************************************************************
M1 1 10 0 0 NSS W=13.2U L=2. 25U
VDS 20 0
```

```
* VGS is positive for nMOS and negative for pMOS

VGS 10 0 5V

* VIDS defines current direction for drain

VIDS 20 1

.DC VDS 0 5 0.05

.PRINT DC I(VIDS)

.PLOT DC I(VIDS)

.WIDTH IN=75 OUT=75

.END
```

Calculation of current-voltage characteristics of .CMOS

Used for the SPICE (Simulation Program with Integrated Circuits Emphasis) simulation program.

```
* CMOS OPERATIONAL AMPLIFIER *
****   INPUT LISTING                      TEMPERATURE =    27.000 DEG C
*****************************************************************************
**************
.MODEL MP1 PMOS (LEVEL=3 TOX=250E-10 VTO=0.55
+ GAMMA=0.38 KP=25.2E-6 NSUB=2E16 THETA=0.163
+VMAX=1E5 FTA=0 DELTA=0 KAPPA=0.8 CGSO=0.65N CGDO=0.65N )
.MODEL MN1 NMOS (LEVEL=3 TOX=250E-10 VTO=0.55
+ GAMMA=0.1 KP=86.8E-6 NSUB=2E16 THETA=0.08
+ VMAX=1E5 FTA=0 DELTA=0 KAPPA=0.8 CGSO=0.42N CGDO= 0.42N )

VDD 10 0 DC 5
VSS 11 0 DC -5
CCR  4 5 10P
CCH  5 0 20P
RIN 16 0 1

* AMPLIFIER
```

```
M1 2 2 10 10 MP1 L = 7.7U W = 8.5U
M2 3 2 10 10 MP1 L = 7.7U W = 8.5U
M3 2 15 1 11 MN1 L = 4.9U W = 25U
M4 3 16 1 11 MN1 L = 4.9U W = 25U
M5 1 6 11 11 MN1 L = 7.9U W = 5U
M6 5 6 11 11 MN1 L = 7.9U W = 19U
M7 5 3 10 10 MP1 L = 7.7U W = 64U
M8 3 9 4 10 MP1 L = 7.7U W = 4U
* VOLTAGE SOURCE
M9 7 7 10 10MP1 L = 7.7U W = 8.5U
M10 6 6 7 10 MP1 L = 7.7U W = 8.5U
M11 6 6 11 11 MN1 L = 7.9U W = 20U
M12 8 8 10 10 MP1 L = 7.7U W = 64U
M13 9 9 8 10 MP1 L = 7.7U W = 50
M14 9 6 11 11 MN1 L = 7.9U W = 19U
```
}Voltage sources

```
VIN 15 0 AC 1

.AC DEC 5 100 100MEG
.PLOT AC VD8 (5) VP (5)
.WIDTH OUT =80
.END
```

1.1.2. *Technologies: conception-aided design*

CMOS operational amplifier

Technology

The technology used is as follows:

– passivation: e_{sio2}: 0.5 – 1 µm;

– N^{++}: heavily doped buried layer → low resistance;

– P^+ zones: insulation wells;

– P^+-N junction in reverse → insulation;

– N^{++} zone: only for "beep",

– $Si_3 N_4$: silicon nitride (mask for implants);

– $Si_3 N_4$: prevents oxide growth (but beaks of parasitic birds at the edge of masks: birds peaks).

Below, we shall resume the basic steps of manufacturing the two preferred devices of microelectronics. Diagrams for this have been given in Volume 1.

Bipolar junction transistor

N_epitaxy collector region (0.1–10 Ω.cm):

$$e_{Nepi} = 2\text{–}15\ \mu m$$

Opening windows are used to create the base and insulation wells. Opening in the base region is used to create the emitter.

Base: a few $\frac{1}{10}$ μμ

N^+: ohmic collector contact

Evaporation of contacts

P-channel MOS:

i) thermal oxidation: $\frac{1}{2}$ μm;

ii) P^+ implementation for drain.

Reoxidation takes place and then so does photo-etching of the oxide.

Gate oxide (dry 0_2 for good dielectric quality); thickness: 0.1–0.15 μm.

Field oxide (FOX: Field OXide; LOCOS: LOCal Oxide on Silicon); very thick: 1–1.5 μm.

Self-aligned gate

I^2: (ionic implantation) = 80 keV; the oxide lets boron through (a very small atom), but the gate stops it, and therefore serves as a natural mask.

Polysilicon gate NMOS inverter

Fabrication process steps:

– $Si_3\,N_4$ is impermeable to O_2;

– boron: P type;

– LOCOS: $Si_3\,N_4$;

– Bird beak: SiO_2;

– threshold voltage adjustment is expressed as: $V_T$adjust;

– phosphorus ⟶ N type.

The I^2 of phosphorus creates source and drain zones and, at the same time, dopes the polysilicone.

The polysilicone "embedded" in the oxide makes it possible to manufacture several levels of interconnections.

Self-aligned polysilicon gate

$$\begin{cases} -\ Creation\ of\ a\ well\ P(10\ \mu\text{m}) \\ -\ Ocide\ growth\ \ (1\ \mu) \end{cases}$$

$$\begin{cases} -\ Window + thermal\ oxide\ (0.1\ \mu\text{m}) \\ -\ \text{Gate oxide} \end{cases}$$

$$\begin{cases} \text{0.6 } \mu \text{ CD } polysilicon\ deposit \\ except\ for\ drain - source\ regions \end{cases}$$

$$\begin{cases} Deposition\ of\ boron - doped\ p\ oxide\ (p, cf.ductility) \\ which\ is\ removed\ from\ the\ (left)\ side\ of\ the \\ \underline{N}MOS; then\ phosphorus - doped\ oxide\ (n) \end{cases}$$

{Furnace → impurities drive in silicon: metallization

– Phase 1: Local oxidation:

a) Thermal growth of a thin layer of silicon oxide.

b) Engraving according to level 1:

- field insulation implantation (B^+ ion);

- annealing of the implantation (annealing, for redistribution).

c) Localized thermal oxidation, followed by the removal of the resin from the nitride and the thin oxide.

– Phase 2: Implementation of depletion:

a) Thin gate oxide growth (≃ 500 Å).

b) Implementation of depletion (P–) according to level 2.

– Phase 3: Adjustment of enhancement transistors:

a) Adjustment implementation (B+) according to level 3.

– Phase 4: Buried contacts or pre-contacts:

a) Etching of the thin oxide according to level 4.

– Phase 5: Gates:

a) Uniform deposition of polycrystalline silicon and etching thereof according to level 5.

b) Source and drain implementation (P *or As*), annealing and oxidation.

– Phase 6: Contacts:

a) Deposition of insulating oxide and etching of contacts according to level 6.

– Phase 7: Metallization:

a) Uniform deposition of aluminum and then etching according to level 7.

– Phase 8: Deposition of final protective oxide and then etching according to level 8. This level only concerns the electric accesses of the circuit: contact and test pads.

Insulated multi-collector outputs

This device operates in reverse mode (the emitter is replaced by the collector and vice versa).

Here, the buried layer serves as the emitter.

It has five manufacturing steps, as follows:

– I^2 N^-;

– I^2 P;

– $-I^2$ N^-;

– contact window;

– metallization.

Power transistor

Double diffusion D.MOS

To ensure a high voltage withstand, we need to create weakly doped areas that are highly resistive, which "take in" the potential differences.

Purpose: Reducing the channel

L< 1 µm

Note: Asymmetric field region → high breakdown voltage.

L: effective length of the channel.

– Highly N-doped substrate is used (→ N-channel MOS).

– Boron implantation is performed (P type).

– Epitaxy (p) from silane (exodiffusion of impurities P from N^- to P): the N^-resting zone will be the field region of the transistor; and the P region (1 µm) will be the channel zone. The drain region is then protected by SiO_2–Si_3N_4–SiO_2 (•) during the growth of the field oxide (*).

– The oxynitride is then removed, and the thick oxide is etched to implant the source region.

Then oxidation occurs. V-shaped anisotropic chemical attack (direction 100>) takes place faster than 111>.

– Growth of the gate oxide; metallization: source, gate and drain.

– Advantages are as follows: the channel lengths obtained are of the order of 1 µm, without using electronic lithography.

– Extending source-drain space loads are essentially vertical → channel shortening effects are negligible.

- P-type diffusion → substrate;

- $I^2\ N^+$ → source.

The polycrystalline silicon gate is in the form of a hexagonal mesh network burying the source regions: at a given surface, the greatest channel density is obtained.

Manufacturing integrated bipolar/CMOS circuits

a) Wafer thickness (wafer, slice, waffle) = 250 µm.

Doping: $10^{16}\,\text{at/cm}^3$.

Low value diffused resistance for collector (20–50 $\mu/\square$).

(As or P because of low diffusivity; large atoms).

Recap: what is resistance per square?

Resistance of a conductive track of length L and width W (a parallelepiped): where L and W characterize the design of the masks, while $R_{\square}$ characterizes the technology. The L/W ratio can be considered as a "number of squares", hence the name of $R_{\square}$: "resistance per square".

$R_{\square} = \rho\ L/S = \rho\ L/(We) = (\rho/e)\ L/(W)$; if L = nW, where n is an integer, then:

$R_{\square} = n^*(\rho/e)$: Ω.

b) Growth of an epitaxial layer (the thickness determines the breakdown voltage): BV_{CE0}, where,

BV is the beakdown voltage.

c) Oxide growth.

Boron diffusion (channel stop: isolate devices from each other by reverse P/N junctions): several hours at 1,200°C (20–40 Ω/□).

d) Manufacturing the base: 100–300 Ω/□ and a depth of 1–3 μ after diffusion.

e) Emitter: 2–10 Ω/□ and a depth of 0.5–2.5 μm after diffusion.

Creation of the collector contact takes place.

f) Manufacture of contacts: additional layer of aluminum over the entire surface and etching of aluminum out of contacts occurs. The electrically active region (in the live active region) is the base portion under the emitter.

Integrated circuit NPN transistor

(Recap: the layout concerns mask drawing).

Keep in mind the existence, at the contacts, of some "resistor and capacitance contact" parasitics.

In an integrated circuit, characteristic parameters of a transistor can be modified by modifying its geometry. For example, to have a high current transistor, the width of the device is increased, but emitter-base, collector-base and collector-substrate capacitances are then increased (compromise, etc.):

Hence, the modified frequency response is obtained.

r_{b1} "ends" and r_{b2} "begins" at the front of the emitter diffusion:

$$\leftarrow r_{b1} = \frac{10}{25}\ 100\ \Omega = 40\ \Omega$$

For r_{b2}, two-dimensional effect occurs.

Crowding effects (overpopulation, defocusing of minority carriers in the base, injected from the emitter, attracted by the base contact(s)).

Injecting carriers from the emitter into the base

At very high currents, the injection takes place at the periphery of the diffusion of the emitter $r_b = r_{b1}$.

For low-noise or high-frequency transistors of low r_b, the periphery of the emitter adjacent to the base contact is maximized. For HF transistors, numerous narrow bands connected by base contacts are used as emitter geometry.

Collector resistance: 3D (see Laplace resolution)

$$R = \frac{eT}{WL} \cdot \frac{Lg\frac{a}{b}}{a-b} \qquad [1.1]$$

where:

– T = thickness of the region;

– e = resistance of the material;

– $\left.\begin{matrix} W = width \\ L = length \end{matrix}\right\}$ of upper rectangle;

– a (respectively, b) = ratio of the width (respectively, length) of the lower rectangle to the width (respectively, length) of the upper rectangle.

Bases are equipotential and current flows vertically. Holes are injected from the emitter, parallel to the surface through the N-type base region, and are picked up by the P-type collector before reaching its contact. The base is more doped than the collector → the desertion/depletion zone extends rather into the collector (more resistive).

The base must be wide enough so that the depletion zone does not reach the collector. When $V_{ce\ max}$ is applied:

$$Wb_{min} = 8\ \mu m \text{ for } V_{ce} \nearrow_0^{40v} \text{ deserted zone (Wb: base width)}$$

It ranges from 6 to 8 μm:

$$t_B(\,transit\ time\ in\ the\ base)\frac{Wb^2}{2\ Dp} = \frac{(8\mu)^2}{2\times 10cm^2/s} = 32\ ns \qquad [1.2]$$

PNP to substrate transistor

One reason for the poor performance of high current side PNPs is the relatively small effective square cross-section of the emitter.

The PNP often operates in class B for output stages → $I_c = 10$ mA.

The substrate is used as a collector.

pb: lightly doped substrate → high series resistance of the collector:

→ If strong currents: large voltage drops in the substrate.

→ Deteriorated performance can occur if no real collector is processed.

Because of the low doping of the N epitaxial material, the series base resistance can be very high if the base contact is far from the base active region. In this structure, the base contact (N^+) can touch the diffusion of the emitter P in order to obtain a base contact diffusion as close as possible to the active base.

Disadvantage: Breakdown voltage of $V_{BC} \simeq 7$ V is used; to increase it, it is necessary to separate the diffusion p from the n^+.

1.1.3. *Resistor technologies*

Diffused resistors

The areas that can be used as resistance are base diffusions, emitter diffusions, the epitaxial layer, the active base region area and the epitaxial layer pinched between the base diffusion and the P-type substrate.

Diffused emitter and base resistors

The diffused base resistor is formed from the P-type base diffusion for the NPNs and is located in an isolated region. The epitaxial region containing this resistor is biased in such a way that the PN junction between this resistor and the epi. layer is reversed. Hence the creation of a contact occurs in the N-type epi., which is connected to the end of the most positive resistor or to a potential that is more positive than any end of the resistor:

$$R = \frac{L}{W} R_{\square} \quad [1.3]$$

where L is the length of the resistor and W is the width.

$R_{\square}$: sheet resistance: 100–200 $\Omega/\square$:

$\longrightarrow R = 50\ \Omega - 50\ \text{k}\Omega$

$(W \geq 5\mu)$

With broadcast emitters, we get $R_{\square}$ values up to 10 $\Omega/\square$.

Base pinch resistance

Pinched zone is found between the collector N and emitter N^+; it can be electrically insulated by reverse biasing the emitter-base and collector-base junctions (in practice, N-type regions are connected to the most positive end of the resistor).

$R_{\square}$= 5–15 k$\Omega/\square$ (interesting because wide range).

Since the N region is lightly doped, it is highly dependent on temperature.

Voltage drop on the resistor is ~6 V, breakdown voltage between the top of the diffused emitter and the base diffusion.

Capacitors of integrated circuits

Capacitor of integrated circuits is given as:

$C = 4.0$ to $60\text{pF}/cm^2$

The breakdown voltage ranges from 60 to 100 V.

The capacitance of the substrate/epi layer junction is negligeable.

High capacitance capacitors (tens of pF) take up a lot of the surface area on the circuit. However, at present, thanks to the design of these circuits, it is possible to reduce the capa. values in order to achieve performances which require large, identical or even better values (see compensation in op. amplifiers). Capacitors can be used by mounting them in shunt, for example, on an amp. Multiplying this capacitance by the gain of the amp. This is the Miller effect.

Two categories are found:

1) P.N junction polarized in reverse → capa. of depletion

- V_{BE} is low ~0.7 V (for V_{BC}, it is stronger but the capa per unit area is lower).

2) MOS capa.

Epitaxial and pinch resistors

To increase the $R_{\square}$, we diffuse P on the top of the epitaxial resistance (behavior identical to a junction FET, JFET).

The pinch-off voltage is a function of the thickness of the epi. and doping.

Limiting the pinch-off resistor to low-level operating voltages does not allow its use in circuits where a low bias current must be derived directly from the supply voltage using a high value resistor. The epitaxial layer itself has a much larger square resistance than the base diffusion, and for this application the epi. layer is often used as resistance. For example, the square resistance of an epitaxial layer with a thickness of 17 μm and a thickness of 5 μcm is equal to:

$$R_{\square} = \frac{1}{q\mu m\, N_d T} = \frac{\ell\, epi}{T} = \frac{R\ \mu\text{cm}}{17\ \mu\text{m}} = 2.9\ \Omega/\square$$

Compatible process for high-performance active devices

Some types of circuits require specific devices such as JFETs for switching and low input current amps, as well as MOSFETs and super-beta NPNs for fast analog circuits. Structures require additional masking steps compared to the conventional manufacturing process.

a) Double diffusion JFET

JFETs can be manufactured on the same surface with bipolar devices, using a different technique mode. If the base region of a standard integrated bipolar circuit is used for the JFET channel, as in the pinch resistor, the pinch voltage is too high because of the channel width of the resulting structure. A narrow-channel double-diffused device can be manufactured on the same surface however, by adding diffusion into the process.

The JFET channel is narrower than the NPN base (the grid is made at the same time as the base: redistribution and pre-deposit of the emitter).

Typically,

– U_p (pinch-off) = 2–5 V;

gate leakage current: 10–100 pA.

This technique is widely used for 3.OPs with FET input stages.

– Pb: breakdown voltage V_{GD} ~ 0.7 V (as V_{RE}).

b) Ion-implanted JFETs

Ion implantation makes it possible to distribute impurities in the channel in a precise manner; the profile is well controlled. Good control of the pinch-off voltage is then obtained (which is not the case with double-diffusion JFETs). In addition, the breakdown voltage V_{GD} may be high insofar as the peak of the doping profile is $\backsimeq 10^6$ at/cm^3.

One way to reduce bias currents is to increase the current gains of the input stages of circuits.

A reduction in base width improves the transport factor ($\propto_T$) of the base and the efficiency of the emitter (γ; current gain increases with the decrease in (W_β):

$$(\propto = \frac{\beta_F}{1+\beta_F} = \propto_T . \gamma \;\; with \; \propto_T = \frac{1}{1+\frac{W_\beta^2}{2\tau_\beta Dn}} \;\; and \; \gamma \; = \frac{1}{1+\frac{D_p \, W_\beta \, W_A}{Dn \, L_p \, N_D}} \qquad [1.4]$$

Thus, current gain can be increased by increasing the diffusion time in the emitter or by decreasing the base width. But an increase in the current gain decreases the breakdown voltage B_{CE0} (BV: breakdown voltage).

Flat C-β breakdown voltage

In practice, a standard transistor and a super-beta (b >>100) are constructed on the same surface by a second diffusion. A very narrow base can lead to piercing (punch through; no more neutral base).

MOS transistor

The MOS transistor shows the following characteristics:

– high switching performance;

– low input current amplifiers;

– very high integration.

NOTE.– Surface condition controls are more draconian than controls for bipolar integrated circuit processes.

Double diffusion PNP transistors

Limiting the frequency response of lateral PNPs limits the performance of analog circuits. We either complicate the "design" of the circuit or the manufacturing process. Double diffusion PNP is a solution: the performance is comparable with the NPN transistor.

Thin film resistor

Diffused resistors have the disadvantage of having a high temperature coefficient, poor tolerance and are insulated junctions (parasitic and photocurrent capa. due to radiation). As a solution, the thin layers are deposited onto the surface.

$$R_{\square}(\Omega/\square)\left|\begin{array}{ccc} (chromium, & tantalum, & Cr-SiO\;) \\ \uparrow & \uparrow & \uparrow \\ 10-1{,}000 & 10-100 & 30-2{,}500 \end{array}\right.$$

CMOS technology: manufacturing process

Substrate:

10^{14}- $10^{15}/\text{cm}^3$

(wells) → $10^{15}/10^{16}$ on the surface after diffusion

→ Si_3N_4 masking (on future active areas)

Nitride is impermeable to O_2.

Insulations

– Bird's beak;

– LOCOS (LOCal Oxidation on Silicon);

– FOX: Field OXide.

I^2 → N channel and P channel: adjustment of surface concentrations, therefore threshold voltages are about 0.1–0 V for N and –2 V for P:

$$V_{tM} = \emptyset ms + 2\emptyset g + \frac{Qb}{Cox} - \frac{Qss}{CO^2}$$

Gate creation (polysilicon)

Previously for depletion MOS, a buried contact was created to connect the gate and source.

Creation of the source and drain zones by I^2 (after etching the oxide at these locations).

Depletion mode–enrichment mode

CVD (chemical vapor deposition) of SiO_2 → reduction of parasitic capa. of interconnection metallizations.

Source and drain diffusion, laterally, creates parasitic capa. with the gate.

The geometry greatly influences electric properties: for example, the transconductance, at constant drain current.

Practical special cases

Layout of diffused resistors

For bulk problems, resistors are folded.

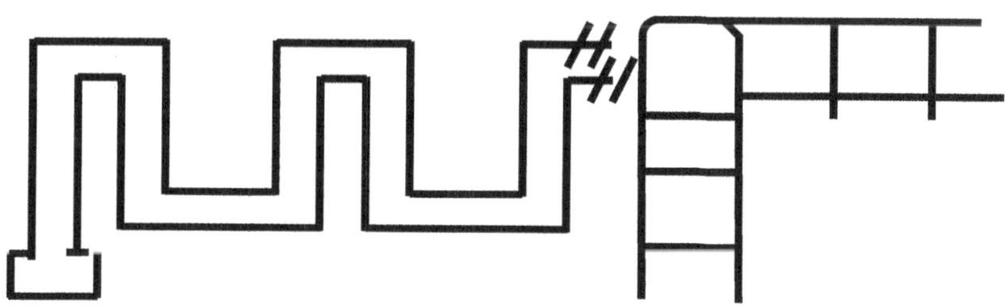

$R = R_{\square} (n_{\square} - 0.44\, n_c)$

n_c: number of elbows
$n_{\square}$: number of squares

Figure 1.1. *A folded resistor*

The resistance of a square at the level of an elbow is equivalent to $0.56R_{\square}$:

$$R_{\square} = \rho/\varepsilon \qquad [1.5]$$

e: thickness of the diffused resistor.

r: mean resistance.

Generally speaking:

$$p = \frac{1}{q(\mu_n N^+ \mu_p P)} \qquad [1.6]$$

Depending on the region, one of the two terms may be ignored.

Switched capacitances

The capacitance of an integrated capacitor is not well controlled (with almost 100% error). Fortunately, there are often capacitance ratios (of the same technology) in transfer functions.

Moreover, it is difficult to control the values of time constants $R \times C$ due to the independence of the technological processes of resistance and capacitance.

Hence, switched capa. is used to simulate resistors, which can have high values.

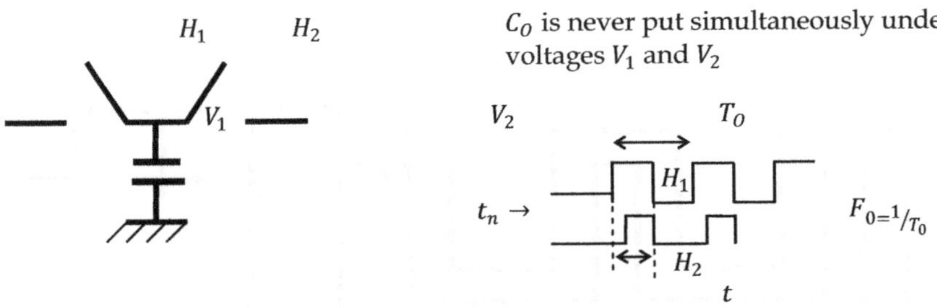

Figure 1.2. *Chronogram of the circuit*

H_y: Capa. charging and discharging are faster than the T_O period:

– at T_n C_O charged to $Q_1 = C_O V_1$

– at T_n C_O charged to $Q_2 = C_O\, V_2 dQ = Q_2 - Q_1 \leftarrow$ at frequency $^1/_{T_O} \rightarrow$ mean current:

$$I = \frac{dQ}{T_O} = \frac{C_O\,(V_2 - V_1)}{T_O} = C_O\,(V_2 - V_1)\,F_O \Rightarrow \frac{1}{C_O F_O} \equiv R_O \qquad [1.7]$$

$$\tau = RC = \frac{C}{C_O F_O} = T_O \frac{C}{C_O}$$

If T_O is well known, accuracy depends on the $\frac{C}{C_O}$ ratio.

1.2. A simple but realistic amplifier circuit: the bipolar junction transistor with a common emitter

In fact, compared with the "academic" case, where the emitter is effectively connected to ground, an emitter retroaction impedance cell $R_E//C_E$ and a base bridge at the input are added.

Here, the purpose is not, once again, to detail all of the formal calculations relating to the existing assembly molds, because they are easily accessible in appropriate literature.

In science, first and foremost, electronics engineers avoid laborious calculations, preferring to adapt, crush configurations, and ultimately, perform simplified, but relatively accurate, calculations. Then, we carry out the more precise calculations via the famous "pooter" (computer).

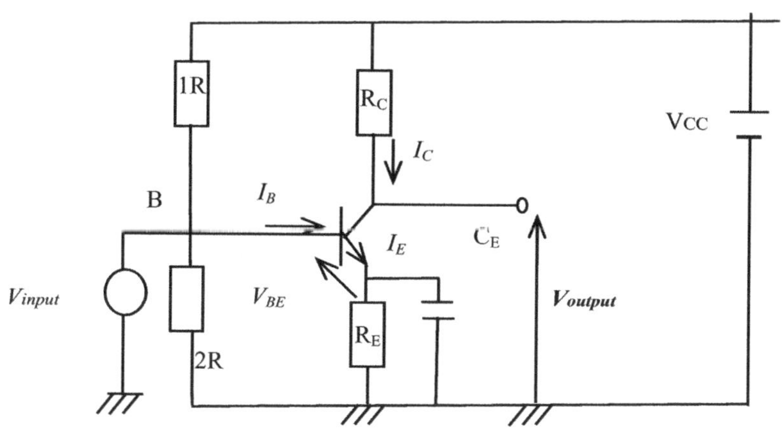

Figure 1.3. *Bipolar junction transistor, NPN, with common emitter, with emitter retraction resistor*

This goes hand in hand with a certain computational "laziness", which gets more and more exalted with an untimely use of the "web".

Here, the base bridge is used to set the base potential, at B, via the power supply.

Introduction of the $R_E///C_E$ in the emitter

R_E performs emitter feedback; Z_E belongs to the input mesh as well as the output mesh. Suppose, for example, an increase Ic due to heating; I_C increases, as well as I^e, V_{EM}, at the R_E terminals too (I_E and Ic >> I_B). Moreover, V_{BM} hardly varies, bias current Ip, passing through R_1 and R_2, being much stronger than Ib entering the base.

Therefore, V_{BE} decreases, and thus I_E or I_C will decrease (the diode "BE" is here forward biased, $I_C = I_E - I_B \cong I_E$ (I_C = Is .exp(V_{BE}/V_T)). Cl: V_{BE} drops and therefore $I_E(I_C)$ decreases.

Feedback

We know that (almost) every signal is a sum of sinusoids of various frequencies and amplitudes; furthermore, the shape of the signal moves from a pure sinusoid, meaning there will be more sinusoids and harmonics. Let's imagine that a harmonic is not a perfect sinusoid, and that a distortion appears at the output; for example a trough, at the level of its maximum. However, this amplifier is mounted as an inverter; a "small" bump will therefore be returned to the input, thus reducing the original hollow.

This acts essentially for the continuous regime. However, the signal will then be badly amplified, even distorted, R_E "puffing" voltage unnecessarily at its terminals, hence the use of a short-circuiting shunting capacitor R_E, in the frequency range of use of the circuit.

1.2.1. *Small signal equivalent schematic of common feedback emitter with base bridge*

In variable signals, for example, alternating signals, the Vcc supply can be likened to a short circuit, because it is constant over time, and therefore of zero variation. Consider the "power supply" as a large capacitor; therefore, it is of very low impedance, tending to a short circuit.

Thus, the collector is grounded, as well as R_1 and R_2, in parallel between the base and the latter.

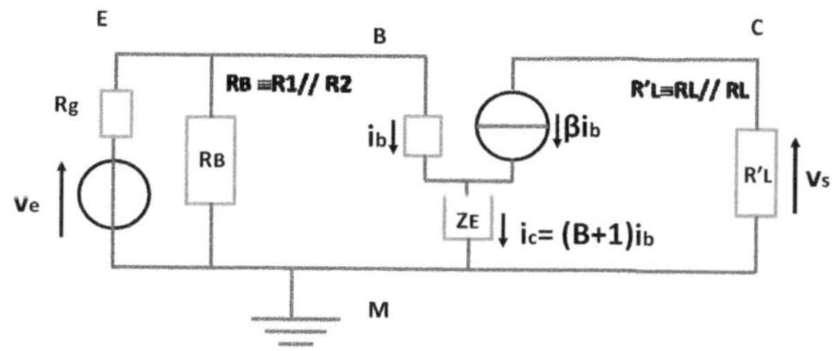

Figure 1.4. *Small signal diagram of the common emitter. For a color version of this figure, see www.iste.co.uk/gontrand/analog2.zip*

1.2.2. *Current gain calculation*

Application of Thévenin's theorem

Let us simplify the diagram before any calculation and consider, for instance, the left side of the circuit.

– R_{TH}

The generator is switched off; its resistance is zero (in fact, this resistance has been output, R_G, with the generator becoming ideal).

$$R_{TH} \equiv R_G \,//\, R_B \qquad [1.8]$$

– E_{TH}

We are dealing with a divider bridge:

$$V_{BM} = \frac{R_B}{R_B + R_G} \qquad [1.9]$$

$$\frac{v_S}{E_{TH}} = \frac{-\beta\ R'_L\ .\ i_b}{R_{TH} + (r_g + (\beta+1)\,R_E).\ i_b} \qquad [1.10]$$

As an alternative, this equation is redirected to:

$$\frac{v_S}{E_{TH}} = \frac{-\beta\ R'_L}{R_{TH} + r_g} \qquad [1.11]$$

If R_E and R_{TH} are of the same order (a few kΩ), and the output voltage β is greater than without R_E decoupling capacitor, typically a 100 ratio.

1.3. Integrated circuit design

Specifications are as follows:

– circuit function;

– its performance;

– its environment (binding, operating conditions: temperature, etc.);

– choosing the best technology.

Architecture

The circuit is broken down into functional blocks. The first being logic: operative part, control part, memory, PLA, connected by internal interconnection buses; the second being analog: operational amplifier: bias circuit, gain stage, output stage.

The integrated circuit to be produced can be studied in real time on a wired model, but this results in a great loss of time for the development thereof. Furthermore, nothing says that mock-up and integrated circuits are identical if everything is "OK" from a functional point of view.

Their behaviors may be very different (problems of parasitic capacitances, input and output resistances of devices). Hence, the interest in numerical simulation is shown.

Conception-aided design (CAD)

In addition, it is possible to create a library of circuits or even functional blocks.

Ground plane

It is also necessary to study the topography of the circuit, to position various blocks with respect to each other, and to estimate the total area, called the ground plane.

The aim being, moreover, to minimize or rather to optimize physical paths traveled by electrical signals, essentially interconnections. One of the methods that is currently becoming widespread is simulated annealing based on the principle of least action (see the traveling salesman problem). Another advantage of computer simulation is to use pre-characterized functional blocks, already modeled and even tested on other circuits.

Layouts and verification

With the layout being completed, it is necessary to check the design rules (Design Rule Checker (DRC)).

The software is then used to ensure that the constraints inherent in the technology are respected; otherwise, the circuit does not work properly, and it will not be made by the smelter.

It should also note the crucial problem of interconnections must be well dimensioned and not be very resistive (e.g. polysilicon wide enough to withstand strong currents). In RF, connection wires have inductance values of the order of nH per millimeter.

Manufacturing/testing

At the end of the design, we have "in hand" a key (historically, a magnetic strip having all of the masking steps) that the silicon wafer will support.

After being manufactured, wafers are tested with the aid of a point machine (it would have been possible, for example, to produce from the aluminum masks, a "point card" adapted to the arrangement of the output pads).

1.4. Current sources

1.4.1. *Simple current sources*

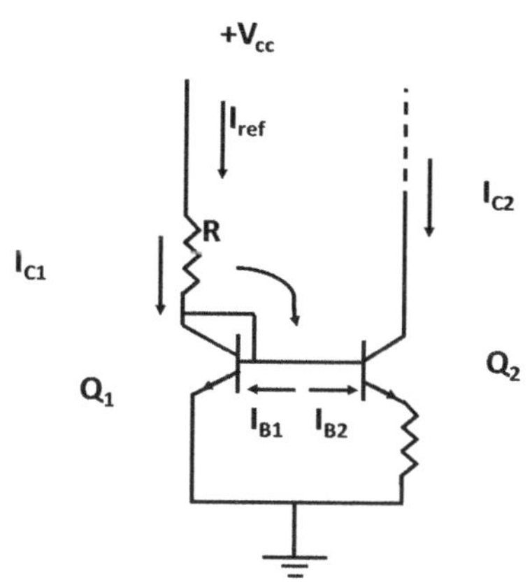

Figure 1.5. *Single current source*

$S_2 = k\ S_1$

$V_{EB1} = V_{EB2}$

$I_{C2} = k\ .I_{C1}$

$I_{ref} = (V_{CC} - 0.6)/R$

$I_{ref} = I_{C1} + I_{B1} + I_{B2}$

$$I_{C_2} = \frac{I_{ref}}{\frac{1}{k}+\frac{1}{\beta}+\frac{1}{\beta k}} \sim kI_{ref} \qquad [1.12]$$

Disadvantage: I_{ref} is a function of R.

1.4.2. *Wildar current source*

$S_2 = k\ S_1$

$I_{S2} = k\ .I_{S1}$

$$I_{S_2} \sim I_S\ exp\frac{V_{BE}}{V_T}$$

$$V_{BE_1} = V_{BE_2} + R_2 I_{C_2}$$

$$V_T\left(Log\frac{I_{C_1}}{I_{S_1}} - Log\frac{I_{C_2}}{I_{S_1}}\right) = R_2 I_{C_2}$$

$$V_T\left(Log\frac{I_{C_1}}{I_{C_2}}\right) = R_2 I_{C_2} \qquad [1.13]$$

1.4.3. *Wilson current source*

Expression of the current delivered by the Wilson source is given as:

$$I_{E3} = I_{E1}$$

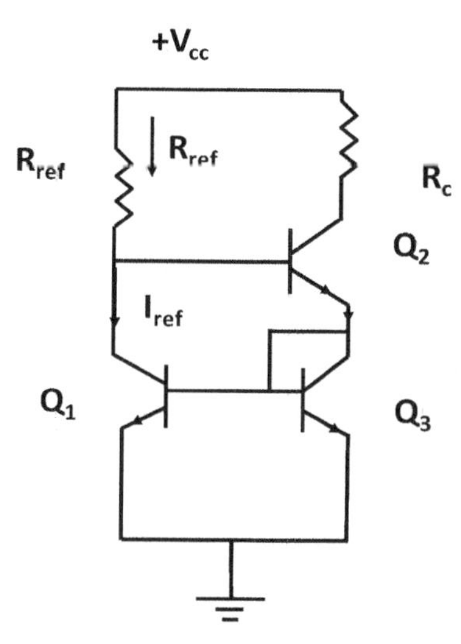

The operating principle is identical to that of the simple source. The reference resistor R_{ref} gives the value of the bias current of Q_1. The current in resistor R_c is then obtained by mirror Q_1, Q_3.

The output resistance is improved due to the feedback produced using Q_2.

An increase in current in R_c causes an increase in the potential V_{BE2}, and therefore in the collector potential of Q_1. The potential difference at the terminals of R_c decreases, which therefore leads to a decrease in the reference current.

– Advantages:

Compared to the simple current source, the Wilson current source delivers a current whose value depends less on the gain b_0 of transistors.

Its output resistance is higher.

– Disadvantages:

It requires three transistors.

Figure 1.6. *Wilson source*

So:

$$I_{REF} - I_{B2} = I_{E2} - (I_{B1} + I_{B3}) \quad [1.14]$$

$$I_{REF} - I_{B2} = I_{E2} + \frac{I_{C_2}}{\beta}\frac{2.I_{C_3}}{\beta}$$

$$I_{REF} - \frac{I_{C2}}{\beta} = I_{c2} + \frac{I_{C2}}{\beta} - 2\,\frac{I_{REF.} - I_{B2}}{\beta} \quad [1.15]$$

$$I_{REF} - \frac{I_{C_2}}{\beta} = I_{c2} + \frac{I_{C_2}}{\beta} - 2\,\frac{I_{REF.} - I_{C2/\beta}}{\beta}$$

$$I_{REF}\left(1 + \frac{2}{\beta}\right) = I_{C_2}\left(\frac{1}{\beta} + \frac{1}{\beta} + \frac{2}{\beta^2}\right)$$

$$I_{C2} = I_{REF}\left(\frac{\beta^2 + 2\beta}{\beta^2 + 2\beta + 2}\right)$$

$$I_{C_2} = I_{Ref}\left(1 - \frac{2}{\beta^2 + 2\beta + 2}\right) \quad [1.16]$$

1.4.3.1. *Equivalent diagram of Wilson source*

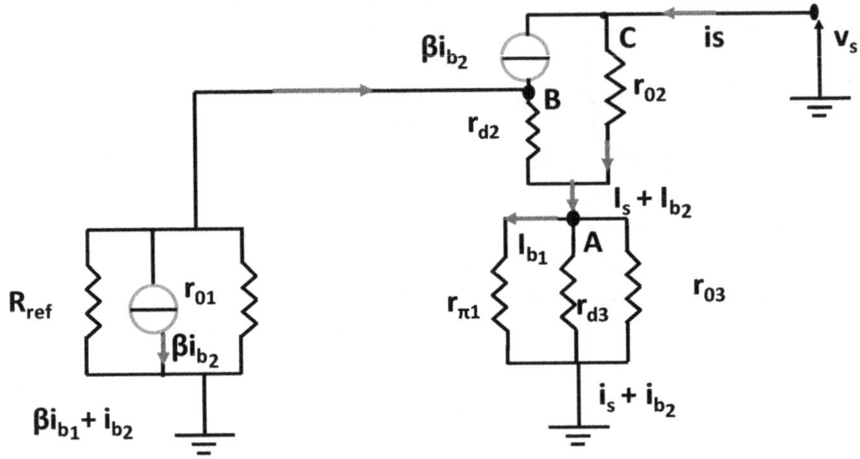

Figure 1.7. *Wilson source: equivalent diagram. For a color version of this figure, see www.iste.co.uk/gontrand/analog2.zip*

1.4.3.2. *Wilson source output resistance*

A voltage source v_s is applied to the output of the source:

$$v_A = r_{d_3}(i_S + i_{b_2})$$

$$v_A - v_S = r_{O_2}(i_S - \beta i_{b_2}) \Rightarrow v_A = v_S - r_{O_2}(i_S - \beta i_{b_2})$$

So:

$$v_S - r_{C_2}(i_S - \beta i_{b_2}) = r_{d_3}(i_S + i_{b_2})$$

$$v_A = r_{d_3}(i_S + i_{b_2}) = r_{\pi_1} i_{b_2}$$

$$v_S - v_A = r_{O_2} i_{O_2} = r_{O_2}(i_S - \beta i_{b_2})$$

$$v_S = v_A + r_{O_2}(i_S - \beta i_{b_2})$$

$$v_S = r_{d_3}(i_S + i_{b_2}) + r_{O_2}(i_S - \beta i_{b_2})$$

$$v_S = (r_{d_3} + r_{O_2}) i_S) + (r_{d_3} - \beta r_{O_2})\, i_{b_2}$$

$$v_B - v_A = r_{d2}(1 + \beta i_{b_2})$$

$$v_B = -Req(\beta i_{b_1} + \beta i_{b_2})$$

$$v_{B-} v_A = r_{d2}(1+\beta) i_{b_2}$$

$$v - v_B = -Req(\beta i_{b_1} + i_{b_2}) + r_{\pi 1} i_{b_1}$$

$$r_{d2}(1+\beta) i_{b_2} = -(Req\beta + r_{\pi 1}) i_{b_1} - Req.i_{b_1}$$

$$i_{b_1} = \frac{r_{d2}(1+\beta)+Req}{\beta Req + r_{\pi 1}} i_{b_2}$$

$$r_{d_3}(i_5 + i_{b_2}) = -r_{\pi 1} \frac{r_{d2}(1+\beta)+Req}{\beta Req + r_{\pi 1}} i_{b_2}$$

$$r_{d_3} i_5 = -r_{d_3} + r_{\pi 1} \frac{r_{d2}(1+\beta)+Req}{\beta Req + r_{\pi 1}} i_{b_2}$$

$$v_S = (r_{d_3} + r_{O_2})\, i_5 - (r_{d_3} - \beta r_{O_2}).r_{d_3} + r_{\pi 1} \frac{r_{d_3}\, i_5}{\frac{r_{d2}(1+\beta)+Req}{\beta Req + r_{\pi 1}}}$$

$$v_S = r_{d_3} + r_{O_2} \frac{\beta r_{O_2} r_d}{r_d(1+\beta) + \frac{r_d \beta + Req}{\beta(Req + r_d}} \qquad [1.17]$$

$$Req = \frac{Vcc}{I} = \frac{Vcc}{Vt} r_d = 15\ \text{V}/0.026\ \text{V} \sim 500\ r_d \rightarrow Req \gg r_d$$

$\beta r_{O_2}/2 >> r_{d_3}$ and r_{O_2}

1.4.4. *Current source and voltage source*

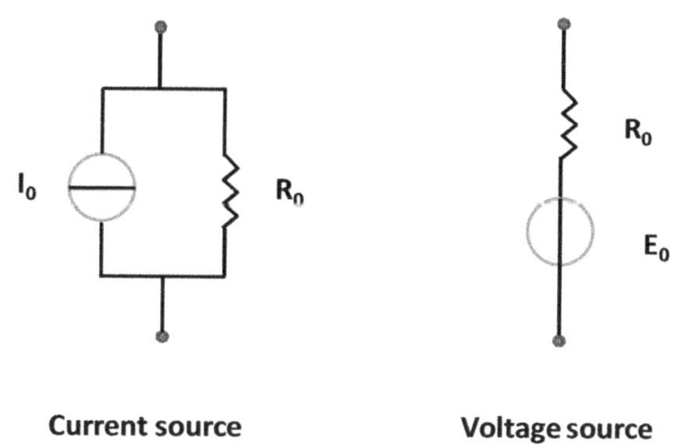

Figure 1.8. *Sources. For a color version of this figure, see www.iste.co.uk/gontrand/analog2.zip*

1.4.5. *Advantages and disadvantages of both sources: one phase and Wildar*

One phase current source

Implementation simplicity:

– It has two transistors and one resistor.

Disadvantages:

– limited output resistance: r_{02} of Q_2;

– high dependence on Vdc;

– poor temperature stabilization;

– cannot deliver low current due to limited R_{ref} values;

– compatible with integration techniques.

Wildar current source

Implementation simplicity:

– It has two transistors and two resistors.

– It can deliver weak currents by adjusting the V_{BE2} bias via R_2.

Disadvantages:

– limited output resistance: r_{02} of Q_2;

– high Vdc dependency;

– poor temperature stabilization.

R_{ref} calculation

It is assumed that $V_{BE1} = 0.6$ V. Hence, the value of R_{ref}:

$$Rref = \frac{V_{CC} - 2\,V_{BE}}{I} = \frac{15 - 1.2}{50\ \mu A} \quad [1.18]$$

$$Rref = 300\ \text{k}\Omega$$

Surface area of Q_1, Q_2 and Q_3

The current in the test transistor is given as:

$$I_c = 1.7 \times 10^{-16} \exp(0.6/0.026) = 1.8 \times 10^{-6}\ \text{A}$$

The surface of the emitters of Q_1, Q_2, and Q_3 is given as:

$S_E = 576*50/1.8 = 16{,}000\ mm^2\ (126\ mm \times 126\ mm)$

IC_{2Min} and IC_{2Min} values

The equivalent diagram of the source is as follows:

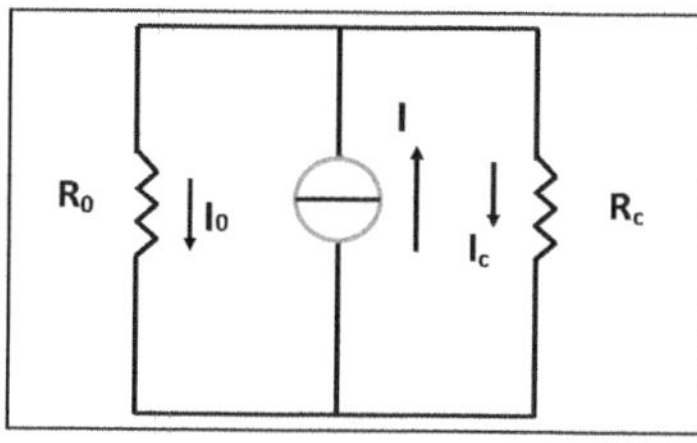

We have: $I = I_0 + I_C$

$R_0 I_0 = R_c I_C$

So:

$R_0(I.I_C) = R_c I_C$

Hence: $Ic = \frac{R_C}{R_0 + R_C} I$

Figure 1.9. *Current divider. For a color version of this figure, see www.iste.co.uk/gontrand/analog2.zip*

a) $R_c = 100kW$:

$I_c^{=4.10}6/^{(4.10}6\ ^{+100.}103)*50\ mA \sim 48.78\ mA = IC_{Min}$

b) $R_c = 1\ kW$:

$I_c = 4 \times 10^{6/}(4.10^6 + 1.10^3) \times 50\ mA \sim 49.99\ mA = IC_{Max}$

Temperature drift coefficient

$$\tau_{CF} = \frac{1}{I_{C2}} \frac{\Delta I_{C2}}{\Delta T}$$

$$\tau_{CF} = \frac{1}{I_{C2}} \frac{\Delta I_{C2}}{\Delta R_{eff}} \cdot \frac{\Delta R_{eff}}{\Delta T}$$

$$\tau_{CF} = \frac{1}{I_{C2}} \frac{\Delta I_{C2}}{\Delta R_{eff}} \cdot \frac{\Delta R_{eff}}{\Delta T}$$

$$\tau_{CF} = \frac{1}{I_{C2}} \frac{\Delta I_{C2}}{\Delta R_{eff}} \cdot \frac{\Delta R_{eff}}{\Delta T}$$

$$I_{C2} = I_{ref} = \frac{V_{CC}}{R_{eff}} ==> \frac{\Delta I_{C2}}{\Delta R_{eff}} = -\frac{V_{CC}}{{R_{eff}}^2}$$

$$\tau_{CF} = \frac{1}{R_{eff}} \cdot \frac{\Delta R_{eff}}{\Delta T} = 2\ 10^{-3}/°\text{C} \qquad [1.19]$$

Study of the pairing of Q_1, Q_2 and Q_3

$$I_{ref} \sim I_{E1} = j_S . S_{E1}\ exp\ (\frac{V_{BE}}{V_T}) \qquad V_{BE} = V_{BE1} = V_{BE3}$$

The current delivered by the source is:

$$I_0 \sim I_{E3} = j_S . S_{E3}\ exp\ (\frac{V_{BE}}{V_T})$$

Therefore, the difference between these two currents is given as:

$$I_{ref} - I_0 = j_S . (S_{E1} - S_{E3})\ exp\left(\frac{V_{BE}}{V_T}\right) = j_S . \Delta S . exp\left(\frac{V_{BE}}{V_T}\right) \qquad [1.20]$$

To represent this current difference, it can be assumed that the surfaces of emitters are equal to S = $(S_{E1} + S_{E2})/2$ and are inserted into the emitter circuit of Q_3, for example, the voltage source corresponding to V_{0S}.

So, we have:

$$I_{ref} - I_0 = j_s\ exp\left(\frac{V_{BE}}{V_T}\right) - j_S . S . exp\left(\frac{V_{BE3} + V_{0S}}{V_T}\right)$$

$$I_{ref} - I_0 = j_s\ exp\left(\frac{V_{BE}}{V_T}\right) - j_S . S . exp\left(\frac{V_{BE3}}{V_T}\right) exp\left(\frac{V_{0S}}{V_T}\right) \qquad [1.21]$$

$$V_{0S} << V_T \rightarrow exp\left(\frac{V_{0S}}{V_T}\right) = 1 + \frac{V_{0S}}{V_T}$$

$$V_{0S} << V_T \rightarrow V_{BE3} = V_{BE3} = V_{BE}$$

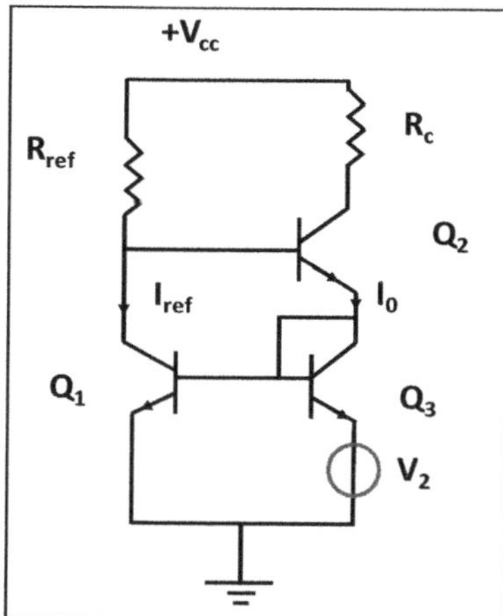

$$I_{ref}-I_0 = j_s.S.exp\left(\frac{V_{BE3}}{V_T}\right)(1 - exp\left(\frac{V_{0S}}{V_T}\right))$$

$$I_{ref}-I_0 = j_s.S.exp\left(\frac{V_{BE3}}{V_T}\right).\left(\frac{V_{0S}}{V_T}\right)$$

So:

$$j_s.\Delta S.exp\left(\frac{V_{BE3}}{V_T}\right) = j_s.S.exp\left(\frac{V_{BE}}{V_T}\right).\left(\frac{V_{0S}}{V_T}\right)$$

that is: $V_{0S} = V_T.\frac{\Delta S}{S}$

Figure 1.10. *Wilson source. For a color version of this figure, see www.iste.co.uk/gontrand/analog2.zip*

I_{C2} current expression

– Consider uneven emitter surfaces

Here:

$$I_{ref}-I_0 = j_s.S.exp\left(\frac{V_{BE3}}{V_T}\right).\left(\frac{V_{0S}}{V_T}\right)$$

$$I_{ref}-I_0 \sim I_{ref}.\left(\frac{V_{0S}}{V_T}\right)$$

$$\left(\frac{I_{ref}-I_0}{I_{ref}}\right) = \frac{V_{0S}}{V_T} \qquad [1.22]$$

– Identical b gains:

$$I_{E1} = I_{E3}$$

$$I_{E1} + I_{B1} = I_{C3} + I_{B3}$$

$$I_{C1} + \frac{I_{C1}}{\beta_1} = I_{E2} - I_{B1}$$

$$I_{C1} + \frac{2I_{C1}}{\beta_1} = I_{B2} + I_{C2}$$

$$(I_{ref}-I_{B2})\left(1+\frac{2}{\beta_1}\right)=I_{C2}(1+\frac{1}{\beta_2})$$

$$(I_{ref}-\frac{I_{C2}}{\beta_2})\left(1+\frac{2}{\beta_1}\right)=I_{C2}(1+\frac{1}{\beta_2})$$

$$I_{ref}(1+\frac{2}{\beta_1})=I_{C2}(1+\frac{1}{\beta_1\beta_2}+\frac{1}{\beta_2}+\frac{1}{\beta_1})$$

$$I_{C2}=I_{ref}\frac{1+2/\beta_1}{1+\frac{1}{\beta_1}+\frac{1}{\beta_2}+\frac{2}{\beta_1\beta_2}})$$

$$I_{C2}=I_{ref}\;(1-\frac{\frac{\beta_2-\beta_1}{\beta_1\beta_2}-\frac{2}{\beta_1\beta_2}}{\frac{\beta_1\beta_2+\beta_1+\beta_2+2}{\beta_1\beta_2}}) \qquad [1.23]$$

We have:

b_2- b_1 >>2

b_2 b_1 >> b_2, b_1, 2

b_2 b_1 = b^2

So:

$$I_{C2}=I_{ref}\;(1-\frac{\Delta\beta}{\beta^2})$$

$$\frac{I_{ref}-I_{C2}}{I_{ref}}=\frac{\Delta\beta}{\beta^2}$$

Different manifold potentials for Q_1 and Q_1

A collector potential variation ΔVc causes the current variation:

$$\Delta I_C=\frac{\Delta V_C}{R_0}$$

Therefore, we have:

$$I_{C2}=I_{ref}+\frac{\Delta V_C}{R_0}$$

$$\Delta V_C = -V_{BE3}$$

$$I_{C2} = I_{ref} - \frac{V_{BE3}}{R_0}$$

But:

$$r_0 = \frac{V_A}{I_{C2}} \sim \frac{V_A}{I_{ref}}$$

So:

$$I_{C2} = I_{ref}(1 - \frac{V_{BE}}{V_A})$$

$$\frac{I_{ref} - I_{C2}}{I_{ref}} = \frac{V_{BE}}{V_A}$$

These three causes of variations are independent and small; we then have:

$$\frac{I_{ref} - I_{C2}}{I_{ref}} = \pm\frac{V_{OS}}{V_T} \pm \frac{\Delta\beta}{\beta^2} - \frac{V_{BE}}{V_A}$$

Or finally:

$$I_{C2} \sim I_{ref}(1 \pm \frac{V_{OS}}{V_T} \pm \frac{\Delta\beta}{\beta^2} - \frac{V_{BE}}{V_A}) \quad [1.24]$$

Numerical evaluation is given as:

$$\frac{I_{C2} - I_{ref}}{I_{ref}} = \frac{2}{\beta^2} = 2.10^{-4}$$

$$\frac{I_{C2} - I_{ref}}{I_{ref}} = \frac{0.6}{200} + \frac{5}{100} + \frac{1}{100} * \frac{20}{100} \sim 5 \times 10^{-2}$$

1.4.6. *Cascode-connected current source*

The name "cascode" was introduced in an article written by Frederick Vinton Hunt and Roger Wayne Hickman in 1939 on the application of voltage stabilizers. They proposed a cascade of two triodes (the first with a common cathode configuration, the second with a common gate) as a replacement for a pentode, and thus the name is assumed to be an abbreviation of "casc(aded triode amplifier having characteristics similar to, but less noisy than, a single pent)ode".

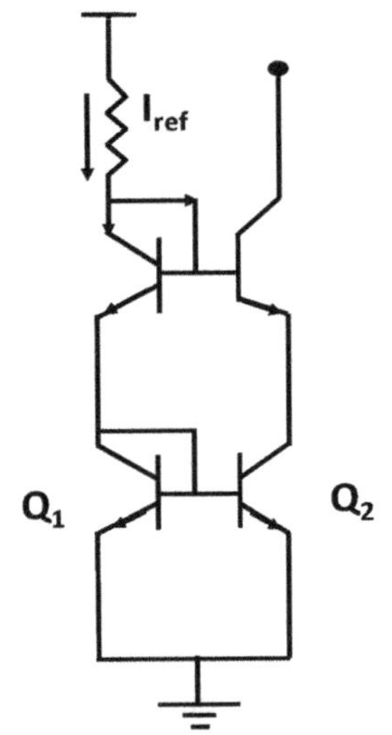

Figure 1.11. *Cascode mounted current source*

1.4.7. *Single current source*

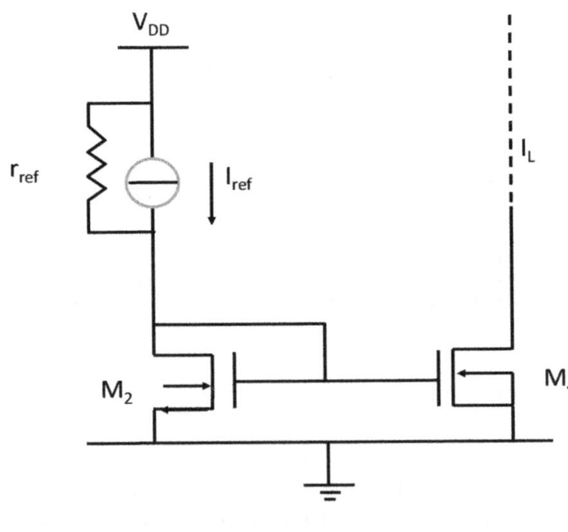

Figure 1.12. *A simple current source. For a color version of this figure, see www.iste.co.uk/gontrand/analog2.zip*

$$\frac{I_L}{I_{Ref}} = \frac{\left(\frac{x}{L}\right)_1}{\left(\frac{x}{L}\right)_2} = \frac{1+\lambda V_{DS1}}{1+\lambda V_{DS2}} \qquad [1.25]$$

If M_1 and M_2 are identical, $I_{ref} = I_L$.

Output resistance

$$R_0 = r_{d_2}.\frac{1+\lambda V_{DS2}}{\lambda\, I_{DS2}}$$

1.4.8. *Improved Wilson current source*

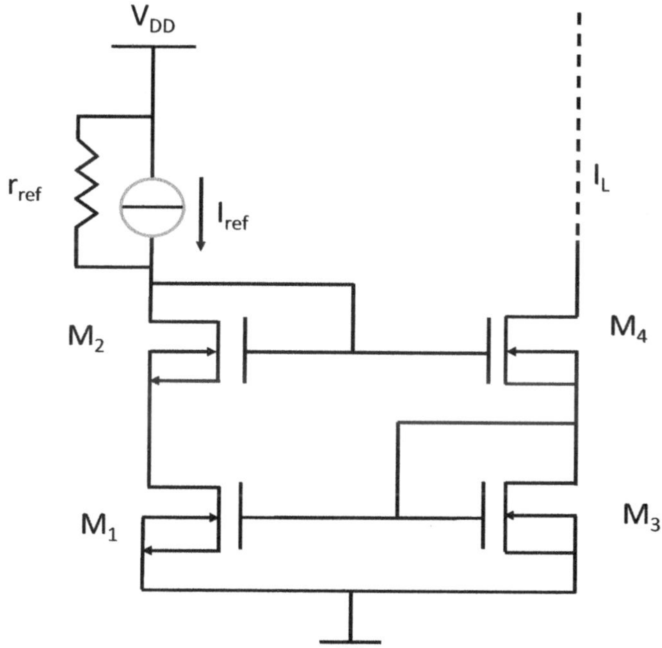

Figure 1.13. *An improved Wilson current source schematic. For a color version of this figure, see www.iste.co.uk/gontrand/analog2.zip*

1.5. A historic circuit: the 741 operational amplifier

Note the size of the area of the MOS *C* capa of 30 pF (compensation: Miller effect, between the output and the input of the gain stage → capa is multiplied by the gain).

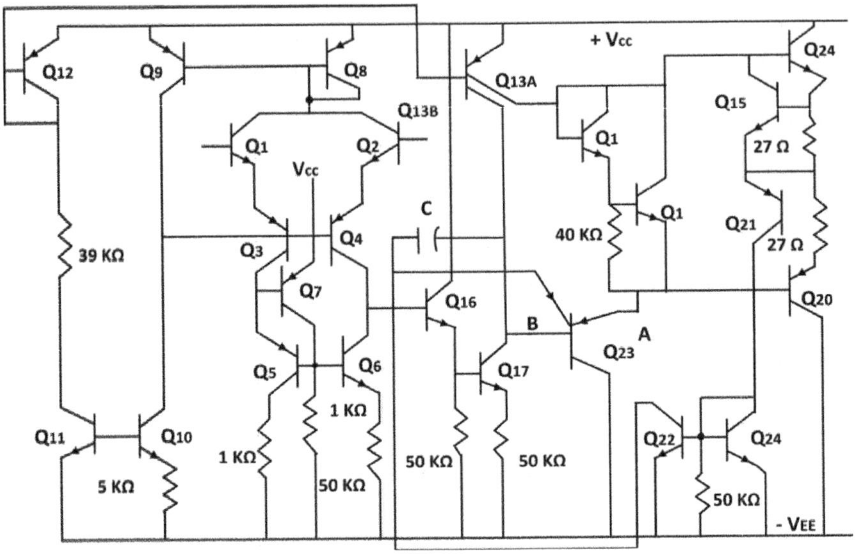

Figure 1.14. *A historical operational amplifier: 741*

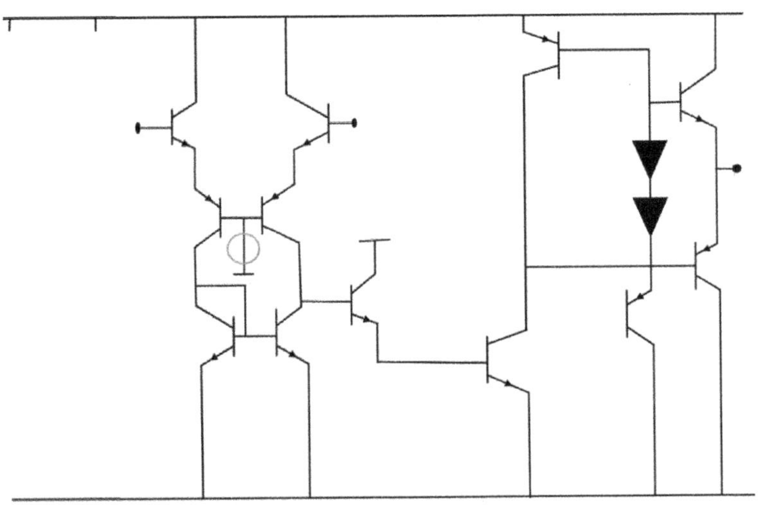

Figure 1.15. *Simplified diagram of 741. For a color version of this figure, see www.iste.co.uk/gontrand/analog2.zip*

– The circuit is an inverting amplifier of gain Av = $-R_1/R_2$, R_2 being located between the input and the lower "lead" of the amplifier, and R_1 between the latter is the output.

– $Z_{M(iller)}$ is between input and output.

– The current flowing in the Miller resistor is: $I_M = (Ve – Ve. (–R_1/R_2))/Z_M$; so:

- the equivalent impedance seen from source Ve (ground is at lead +) is $Ve/I_M – Z_M/(1 – Av)$;

- impedance Zeq seen from the source Ve therefore appears as having a value Zm/Av (1 is neglected in front of Av) times lower.

– Here, Zm = 1/jCω; result: the "capa"; view from the inlet is almost Av€larger than the effectively welded shunt capa.

This makes it possible to avoid capacitors of too large surface area, which is unacceptable for integrated circuits.

Q_1, Q_2: follower emitters (common collectors). Then, these devices present high input impedance.

This circuit presents a low input current.

They control Q_3 and Q_4 (common NPN base).

Q_5, Q_6: active charge of Q_3 and Q_4.

1.5.1. *Active charge*

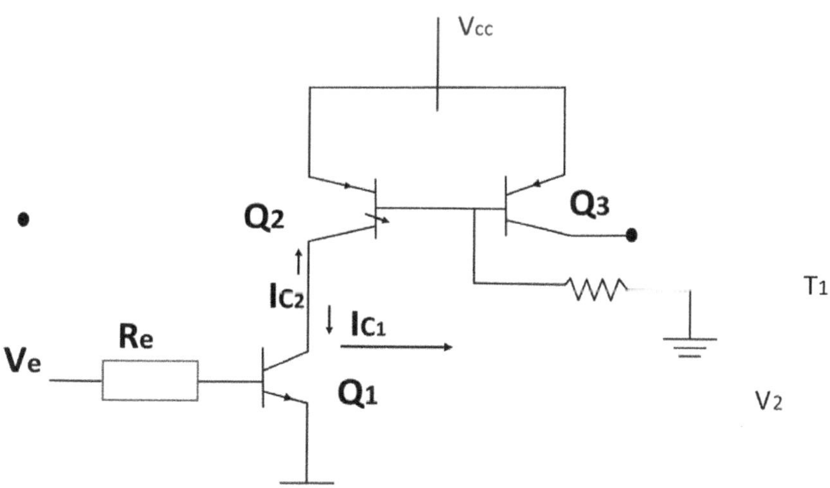

Figure 1.16. *Active charge. For a color version of this figure, see www.iste.co.uk/gontrand/analog2.zip*

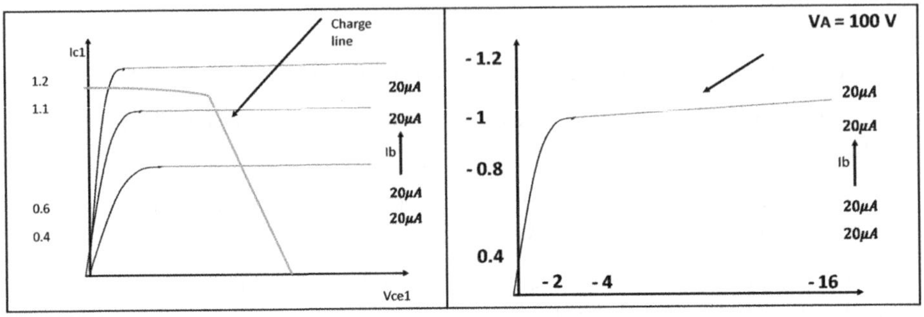

Figure 1.17. *Characteristics of the active charge. For a color version of this figure, see www.iste.co.uk/gontrand/analog2.zip*

1) $I_{C2} = -1$ mA, so: I_{C1} and $V_{CE2} = 15$ V – 1.5 V = 14 V.

2) $I_{C1} = 0$ A, so $V_{CE2} = 0$ V ➔ $I_{C1} = 0$ mA, so: $V_{CE1} = 15$ V.

3) $I_{C2} = -1.1$ mA ➔ $I_{C1} = 1.2$ mA, so : $V_{CE1} = 15$ V – 15 V = 0V.

Therefore:

Charge line $r_0 = V_A/IC_2 = 100$ V/1 mA = 100 kW

NOTE.– The charge line strikes the V_{CE1} at V_A (–100 V), because $I_{C2} = 0$ if $V_A = \_100V$ <==> = 0.

The output resistance of the source is equivalent to the small signal collector output resistance.

1.5.2. *741 description*

a) The six transistors provide a differential input relatively insensitive to common mode voltages of high input impedance and having a voltage gain (noise and offset voltage brought back to the input are divided by this gain).

b) Level shifter:

Q_3 and Q_4 are PNPs: indispensable for shifting the DC level to the negative ($V_{c_{Q_3}}$ and $V_{c_{Q_4}} \simeq -V_{EE}$).

NOTE.– PNP transistors have a poor frequency response (see base, N type, weakly doped) and a low β/NPN.

c) Differential input-single output conversion.

We could take a single output of the differential pair, with problem of high sensitivity of the input voltages in common mode → use of an active charge circuit (Q_5, Q_6).

Q_{16}: Emitter follower: adapter reduces the charge effect of Q_{17} on the output of the active charging stage.

Q_{17}: Common emitter amplifier with active charge.

Q_{13B}: Large voltage gain.

Q_{23}: Emitter follower: prevents the output stage from charging the output of the gain stage.

Q_{14}, Q_{20}: Class AB output stage.

Q_{13}: Multicollector lateral PNP.

I_S: Saturation current when the two collectors are connected.

1.5.3. *Continuous analysis*

We deal here with the calculation of the quiescent currents and voltages.

Pb: Saturation (typically if 0.1 mV offset voltage). At the input, a voltage gain of 10^5, and a zero input voltage is highlighted for small variations in the β or output resistances of the stages (grounded inputs).

Cl: first hypothesis. The circuit is included in a feedback loop "forcing" the output to a fixed voltage, typically 0.

Second simplifying hypothesis: The output resistance of transistors barely has an effect on direct currents flowing in the circuit (error of 10–20% in the calculation of these currents).

NOTE.– In the AC analysis, the output resistance must be taken into account, because it strongly affects the gain for active charge amplifiers.

Determination of bias currents in the input stage

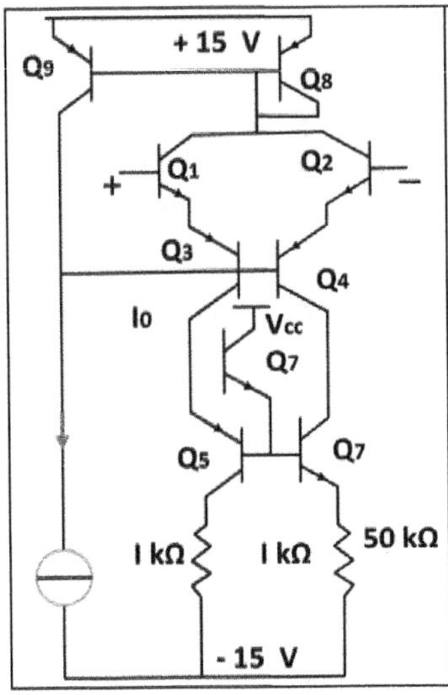

– The continuous base currents of the NPNs (see β of several hundred) but not of the PNPs (much weaker β) are ignored.

$$I_A = -(I_{c9} + I_{B9} + I_{B8})$$

$$= -\left(I_{c9} + \frac{I_{c9}}{\beta_{Q9}} + \frac{I_{c8}}{\beta_{Q8}}\right)$$

$$= -I_{c9}\left(1 + \frac{2}{\beta_{pnp}}\right)$$

Figure 1.18. *Input: polarization. For a color version of this figure, see www.iste.co.uk/gontrand/analog2.zip*

Calculation of bias currents Q_{11} and $Q_{13\ AB}$:

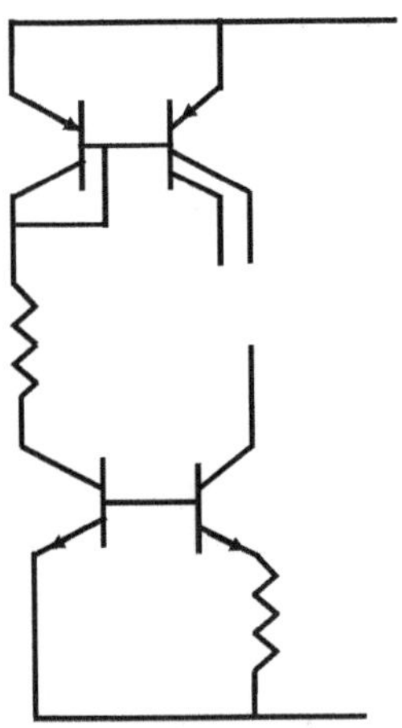

Figure 1.19. *741 polarization source*

Wildar current source (Q_{10}, Q_{11}):

$$\rightarrow V_T Log \frac{Iref}{I_1} = R_4 I_1 \quad [1.26]$$

Graphic or digital solution:

$I_1 = 19\ \mu A$

In addition, $I_2 = \frac{3}{4}\ I_{ref}$ and $I_3 = \frac{1}{4}\ I_{ref}$

that is, $I_2 = 0.55$ mA and $I_3 = 0.18$ mA

The circuit can then be simplified as follows:

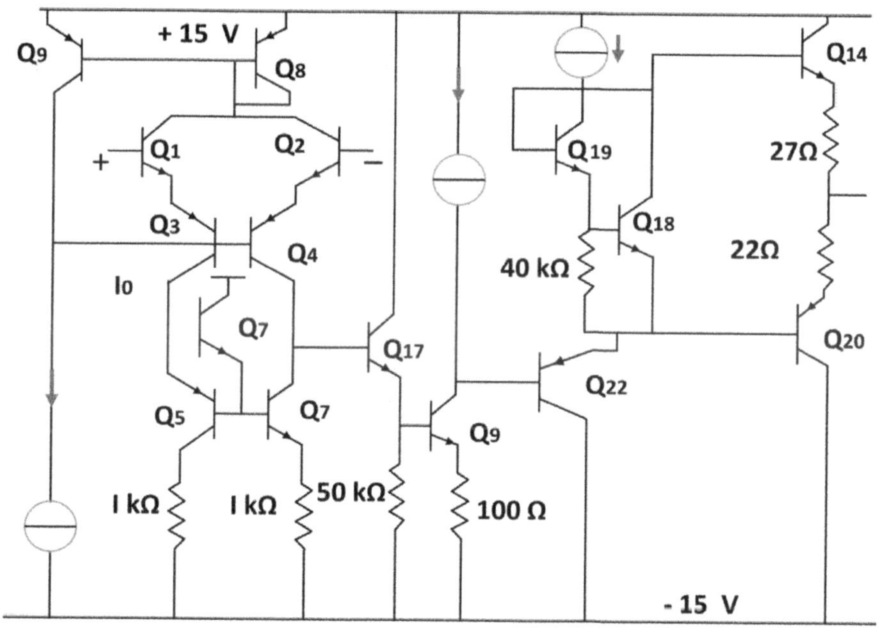

Figure 1.20. *741 polarization currents. For a color version of this figure, see www.iste.co.uk/gontrand/analog2.zip*

$$I_{E3} = I_{E4} = \frac{I_A}{2} + I_{B1,2} = -\frac{I_{C9}}{2}(1 + \frac{2}{\beta pnp})$$

But $I_1 = I_{B3} + I_{B4}.\ I_{C9}$

$$= \frac{I_{E3}}{1+\beta pnp} + \frac{\mathrm{I}_{E4}}{1+\beta pnp} - I_{C9}$$

$$= - I_{C9}\left[1 + \left(\frac{1+\frac{2}{\beta pnp}}{1+\beta pnp}\right)\right] \qquad [1.27]$$

$$= 19\ \mu\text{A}$$

$$I_A = -I_{C9}.\left(1+\frac{2}{\beta pnp}\right) = \frac{19\mu A}{\left[1+\left(\frac{1+\frac{2}{\beta pnp}}{1+\beta pnp}\right)\right]}.\left(1+\frac{2}{\beta pnp}\right)$$

$$I_A \simeq 19\ \mu A\left(1+\frac{1}{\beta pnp}\right) \qquad [1.28]$$

cl: If βpnp is large enough:

$$\boxed{I_{C1} = I_{C2} = 9.5\ \mu A}$$

$$\boxed{(I_{C3} = I_{C4}) = (I_{C5} = I_{C6}) \simeq 9.5\ \mu A}$$

↑

If base currents are ignored

The bias feedback loop formed by Q_8 and Q_9 thus stabilizes the bias current in each input transistor (Q_1, Q_2) at about half the collector current of Q_{10}.

Calculation of the current in the collector of Q_7:

$$I_{E7} = I_{\beta 5} + I_{\beta 6} + I_{50k\Omega}$$

$$V_{50k\Omega} = V_{\beta E5}(V_{\beta E6}) + V_{1k\Omega}$$

with $I_S = 10^{-14}$A and I_{C5} (or I_{C6}) = 9.5 μA → $V_{\beta E5(6)}$ = 537 mV:

and $V_{1k\Omega}$ = 9.5 mA

$$\rightarrow I_{50k\Omega} = \frac{547\ \text{mV}}{50\ \text{k}\Omega} = 11\ \mu\text{A}$$

So $\boxed{I_{C7} = 11\ \mu\text{A}}$

Let us consider Q_{16}

$$V_{\beta 17} = V_{\beta E17} + V_{100\Omega}$$

But $I_{C17} = 550\ \mu A$ (current supplied by $Q_{13\beta}$)

$$\rightarrow V_{\beta 17} = (550\ \mu\text{A})(100\ \Omega) + V_T \log \frac{550 \times 10^{-6} A}{250}$$

$$= 55\ \text{mV} + 642\ \text{mV}$$

$$\boxed{V_{\beta 17} = 697\ \text{mV}} \quad \rightarrow \quad I_{\beta 17} = \frac{550\ \mu\text{A}}{250} \quad \rightarrow \quad \boxed{I_{\beta 17} = 2.2\ \mu\text{A}}$$

So $I_{C16} = I_{50k\Omega} + I_{\beta 17}$

$$= \frac{697\ \text{mV}}{50\ \text{k}\Omega} + 2.2\ \mu\text{A}$$

$$\boxed{I_{C16} = 16\ \mu\text{A}}$$

– Output stage ($V_{out} = 0$ for the calculation of the bias current).

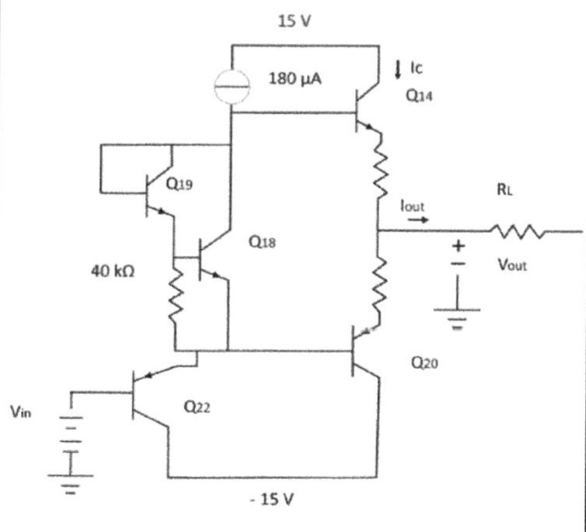

Negligible base currents
→ $I_{C23} = I_{E23} = 180\ \mu$A.
The Q_{18}, Q_{19} pair (Darlington), take up little space, making it possible to overcome the problem of the distortion of the output characteristic.
It is sufficient for the potential difference, at the terminals of the two bases, to be approximately 1.2 V, that is, twice the threshold voltage of a diode; the two transistors conduct at rest, even if only slightly. Moreover, this pair makes it possible to prevent this characteristic from passing through the origin; in fact, with two simple diodes in series (see below), a shift would persist; the output characteristic then goes to zero only when $V_{in} = -V_{BE18}$.

Figure 1.21. *741 output. For a color version of this figure, see www.iste.co.uk/gontrand/analog2.zip*

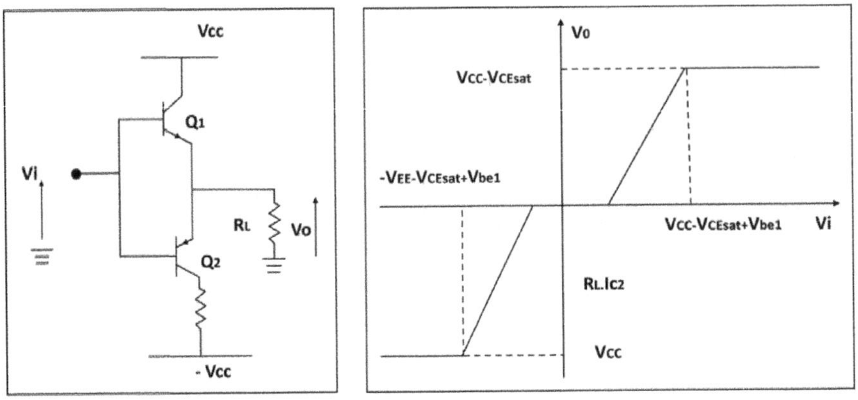

Figure 1.22. *(a) Push-pull. (b) Voltage gain*

– Calculation of bias currents (DC) in Q_{14}, Q_{18}, Q_{20} and Q_{19}.

Zero output voltage and current (considering the feedback):

$$\rightarrow I_{C14} \simeq I_{C20}$$

$$I_{C15} \simeq \frac{0.6\ \text{V}}{40\ \text{k}\Omega} \quad \rightarrow \boxed{I_{C19} = 15\ \mu\text{A}} \quad (I_{\beta 18} = 0)$$

If $I_{\beta 18} = 0$ then $I_{C18} = 180 - 15\ \mu\text{A} = 165\ \mu\text{A}$

And so:

$$I_{C15} = \frac{165\ \mu\text{A}}{\beta_F} + \frac{V_T\ Log\frac{165\ \mu\text{A}}{10^{-14} A}}{40\ \text{k}\Omega} \quad [1.29]$$

$I_{C19} = 15.6\ \mu\text{A}$: to be compared with the 15 μA above:

$$I_{C14} = -I_{C20} = \sqrt{I_{C19}\, I_{C18}} \cdot \sqrt{\frac{I_{S14}\, I_{S20}}{I_{S18}\, I_{S19}}} \quad [1.30]$$

The voltages $V_{27\Omega}$ in series with the emitters of Q_{14} and Q_{20} (and the output of the push-pull) are ignored. The quiescent currents that flow in the output transistors depend on the value of their I_S, which depends on their geometric structure.

These two transistors must withstand high currents while maintaining good β gains when a low load resistance is output; therefore, they have a larger gauge than the other transistors of the circuit.

Although their geometry depends on the "foundry", their I_S is typically three times larger than that of transistors with small geometry, that is:

$$I_{C14} = -I_{C20} = \sqrt{(15.6\ \mu A)(165\ \mu A) \times 3}$$

$$\boxed{I_{C14} = -I_{C20} = 152\ \mu A}$$

Protection circuits

In practice, several devices are active only under overload conditions.

– Q_{15} operates only when the voltage at the terminals of the 27 Ω exceeds ≃600 mV, that is, when the output current is ≃20 mA. When Q_{15} is conducting, the base current of Q_{14} is limited.

→ Q_{15}: short-circuit protection.

– Similarly, Q_{21}, Q_{22}, and Q_{24} protect from short circuits.

– Also the additional emitter of Q_{23} avoids the excess dissipation that could occur if Q_{17} became saturated.

Suppose that the inverting input has been set to a more positive potential than the non-inverting input so that Q_1 is blocked. The current in the base of Q_{16} would be 20 µA.

Without the additional emitter of Q_{23}, this current would be amplified by a β which can reach a value of 1,000 and would reach 20 mA.

This would circulate in the base of Q_{17} (which could saturate) until reaching $-V_{EE}$ through the 100 Ω of the Q_{17} emitter. Hence, a power dissipation of 20 mA × 30 V ≃ 600 W in Q_{16}. Therefore, this emitter prevents Q_{17} from saturating by diverting the base current from Q_{16} when the V_{CB} of Q_{17} reaches 0.

This prevents high current conditions that would damage Q_{16}.

1.5.4. *Analysis of 741 small signals*

Consider the circuit in its three stages:

– input stage;

– gain stage;

– output stage.

The purpose is the calculation of the input, output and transconductance resistance of each stage.

1.5.4.1. *Input resistance*

Consider the alternating diagram of the following input stage:

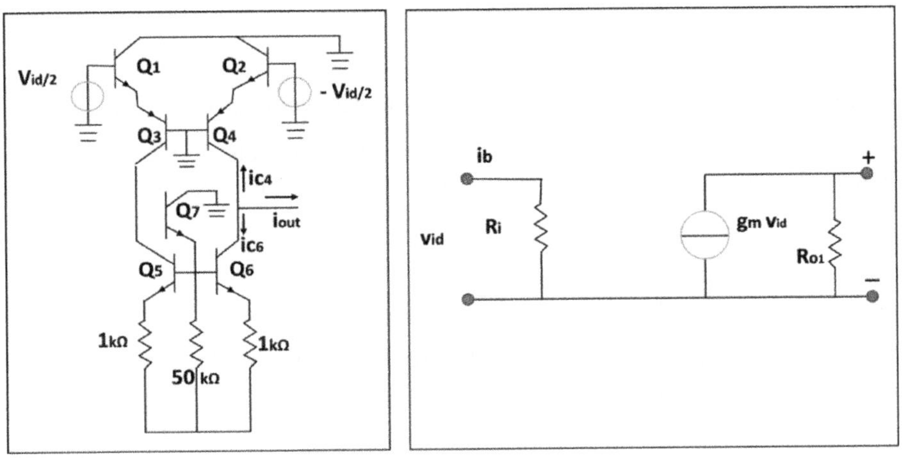

a) b)

Figure 1.23. *Input resistance. For a color version of this figure, see www.iste.co.uk/gontrand/analog2.zip*

PNP bases are grounded (for AC).

Furthermore:

$$r_{\pi 16} = \frac{\beta_0}{g_{m16}} = \frac{\beta_0}{I_{C16}/V_T} = \frac{250}{\frac{16\mu A}{26 mV}}$$

And:

$$r_{\pi 16} = \frac{250}{\frac{550\ \mu A}{26\ \text{mV}}} = 12\ \text{k}\Omega$$

Thus:

$$Ri_2 = 5.7\ \text{M}\Omega$$

1.5.4.2. *Calculation of transconductance* G_m

Short-circuited output

Calculation of the output current resulting from the differential input.

i) No signal flows through the output resistors of:

Q_4 and Q_6 .

ii) The active charging circuit produces a current i_{C6} such that $i_{C6} = -i_{C3}$ (cf Q_3):

Hence, $i_{out} = -(i_{C4} - i_{C3})$

The collector resistance of Q_3 is small/$r_o(output)$ of Q_3. The output resistances of devices are then ignored.

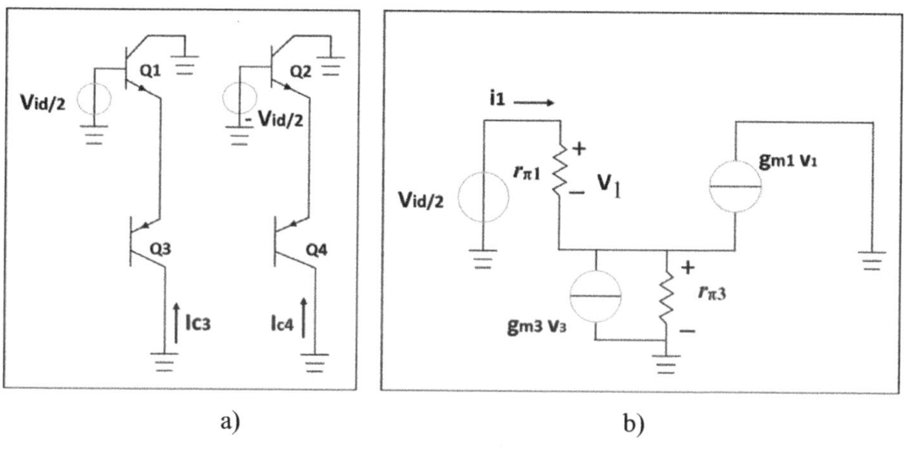

Figure 1.24. *(a) Toward the calculation of* Gm_1 *transconductance; (b) small signal diagram. For a color version of this figure, see www.iste.co.uk/gontrand/analog2.zip*

For a half-circuit:

$$\frac{v_{id}}{2} = v_1 + v_3$$

and:

$$v_1 g_{m1}\left(1 + \frac{1}{\beta_{o1}}\right) = v_3 g_{m3}\left(1 + \frac{1}{\beta_{o3}}\right)$$

with:

$$\beta_{o1} = \beta_o \text{ of } Q_1$$

$$\beta_{o3} = \beta_o \text{ of } Q_3$$

Hence $\frac{v_{id}}{2} = \left[\frac{g_{m3}(1+\frac{1}{\beta_{o3}})}{g_{m1}(1+\frac{1}{\beta_{o1}})} + 1\right]$ [1.31]

Since $|I_{C1}| = |I_{C3}|$, then $g_{m1} = g_{m3}$

Assuming further that β_{o1} and $\beta_{o3} \gg 1$, then

$$v_3 = \frac{v_{id}}{4}$$

and:

$$I_{C3} = \frac{-g_{m3}\, v_{id}}{4}$$

Due to the symmetry of the circuit:

$$I_{C4} = +\frac{g_{m3}\, v_{id}}{4}$$

Hence, according to the penultimate figure:

$$i_{out} = -i_{C4} + i_{C3} = -\frac{g_{m3}\, v_{id}}{2}$$

From where G_{m1} (transconductance of the input stage) $= -\frac{i_{out}}{v_{id}}$

$$= \frac{g_{m1}}{2} = \frac{1}{5.4\ k\Omega} = 18 \text{ mS}$$

1.5.4.2.1. Input resistance: R_i

In addition, the differential input resistance can be calculated (see Figure 1.23):

$$Req = \frac{1}{g_{m3}(1+\frac{1}{\beta_{o3}})} \quad [1.32]$$

The resistance appears in the emitter of Q_1 and knowing that:

$$R_i = r_\pi + R_E(\beta + 1)$$

we have:

$$\frac{\frac{v_{id}}{2}}{i_1} + r_{\pi 1} + Req(\beta_{o1} + 1) = \left[r_{\pi 1} + \frac{(\beta_{o1} + 1)}{g_{m3}(1 + \frac{1}{\beta_{o3}})} \right] \quad [1.33]$$

If $\beta_{o3} \gg 1$ and $g_{m1} = g_{m3}$:

$$\frac{v_{id}}{2} = \left(r_{\pi 1} + \frac{\beta_{o1}}{g_{m3}} \right) i_1 = (r_{\pi 1} + r_{\pi 1})\, i_1$$

$$\frac{v_{id}}{i_1} = R_i\, d = 4\, r_{\pi 1} = 2.7\, M_\Omega \qquad (with\ \beta_o = 250)$$

The input resistance of the amplifier is therefore four times greater than the input resistance of one of the input transistors. In this calculation, we neglected the fact that when v_{id} changes, the output voltage changes and produces a feedback to the input through the output resistance of Q_4.

This effect produces a difference between the input resistances at the two input terminals.

1.5.4.2.2. Calculation of the output resistance R_{o1} of the input stage

The input voltage is set to zero.

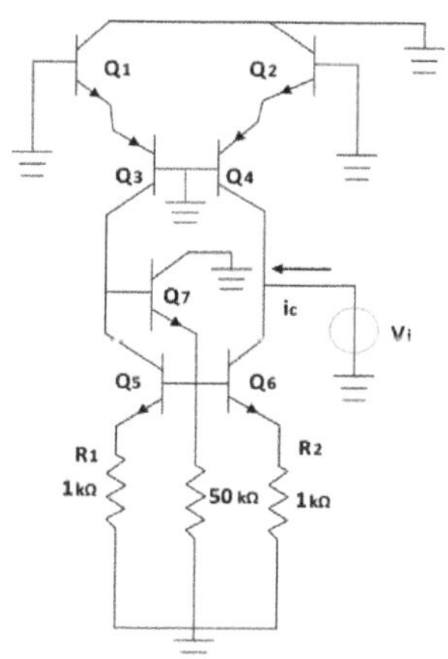

Figure 1.25. *Calculation of the output resistance. For a color version of this figure, see www.iste.co.uk/gontrand/analog2.zip*

The Thévenin-equivalent resistance (at the base of Q_6) is small compared to the r_π of Q_6; therefore, the error is small if it is assumed that the base of Q_6 is grounded. Thus, the only output resistors that are significant for determining R_{o1} are those of Q_4 and Q_6, whereas all of the collector outputs are at points of low impedance. The circuit is not balanced so that PNP transistor bases are not at virtual ground potentials. However, this hardly changes the results.

So:

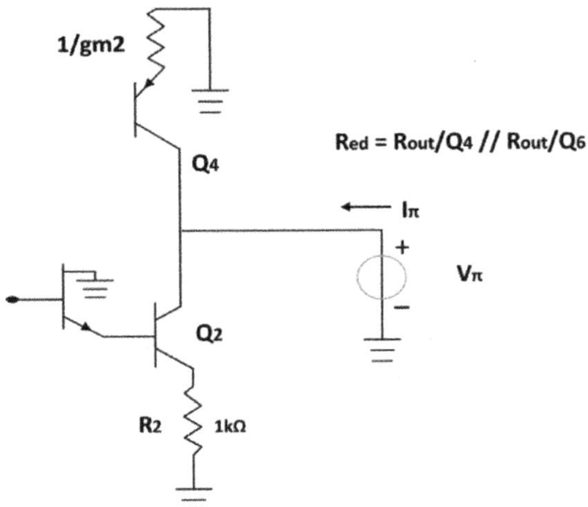

Figure 1.26. *Input resistor. For a color version of this figure, see www.iste.co.uk/gontrand/analog2.zip*

The resistance in the Q_6 emitter is 1 kΩ, while that of the Q_4 is the r_e of Q_2.

$$R_{ed} = (R_{out}/Q_4//R_{out}/Q_6)$$

The output resistance of current source transistors with an emitter resistance is given by:

$$R_o \simeq r_o \left(\frac{1+g_{m2} R_2}{1+\frac{g_{m2} R_2}{\beta_o}} \right) \quad [1.34]$$

And thus:

$$R_{o1} = \left\{ r_{o4} \times \left[\frac{1+g_{m4}\left(\frac{1}{g_{m2}}\right)}{1+\frac{\frac{g_{m4}}{g_{m2}}}{\beta_{o4}}} \right] \right\} // \left\{ r_{o6} \times \left[\frac{1+(g_{m6})\,(1k\Omega)}{1+\frac{(g_{m6})\,(1k\Omega)}{\beta_{o6}}} \right] \right\} \quad [1.35]$$

If β_{o4} and $\beta_{o6} \gg 1$, then

$$R_{o1} = 2\, r_{o4} \,//\, 1.36\, r_{o6}$$

For $\eta_{NPN} - 2 \times 10^{-4}$, $I_C = 9.5\ \mu A$, $\eta_{PNP} - 5 \times 10^{-5}$

We find $R_{o1} = 7\ \text{M}\Omega$.

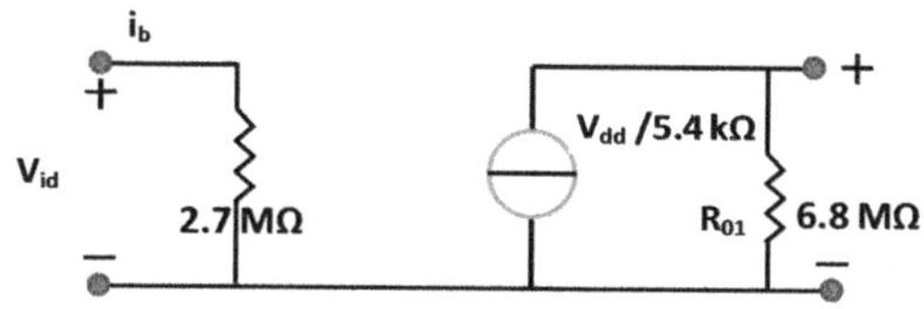

Figure 1.27. *Input stage. For a color version of this figure, see www.iste.co.uk/gontrand/analog2.zip*

Let us look at the second stage.

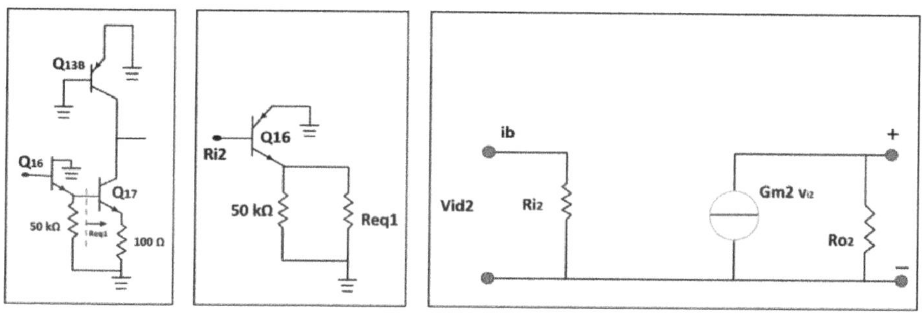

Figure 1.28. *Central gain amplifier. For a color version of this figure, see www.iste.co.uk/gontrand/analog2.zip*

We go to perform some calculations for input resistance, transconductance and output resistance.

1.5.4.3. *Input resistance: Re_2*

The Thévenin equivalent resistance seen between the base of Q_{17} and the ground: Req_1 (Figure 1.28(b)).

Q_{17} common emitter $\rightarrow$ $Req_1 = [r_{\pi 17} + (\beta_{017} + 1).\, 100\ \Omega]$

The effects of the output resistance of Q_{17} have been ignored; the positive feedback due to the fact that it only slightly decreases the input resistance.

Hence the simplified diagram in (Figure 1.27):

$$\rightarrow R_{i2} = r_{\pi 16} + (\beta_{o16} + 1)\ (50\ \text{k}\Omega\ //\ Req_1)$$

$$= r_{\pi 16} + (\beta_{o16} + 1)\{50\ k\Omega\ //\ [r_{\pi 17} + (\beta_{o17} + 1) \times 100\ \Omega]\}$$

Suppose $\beta_o = 250$

In addition, $I_{C16} = 16\ \mu$A and $I_{C17} = 550\ \mu$A:

$$\Rightarrow \quad r_{\pi 16} = \frac{\beta_o}{g_{m16}} = \frac{\beta_o}{\frac{I_{C16}}{V_T}} = \frac{250}{\frac{16\ \mu\text{A}}{26\ \text{mV}}} = 400\ \text{k}\Omega \qquad [1.36]$$

and:

$$r_{\pi 17} = \frac{250}{\frac{556\mu\text{A}}{26\text{mV}}} = 12\ \text{k}\Omega \qquad [1.37]$$

Therefore:

$$R_{i2} = 400\ \text{k}\Omega + (250{+}1)\ \{50\ k\Omega//(12\ \text{k}\Omega + 251 \times 100\ \Omega)\}$$

$$= 405 + 251\ (50\ \text{k}\Omega//\ 37\ \text{k}\Omega)$$

$$\sim 5.5\ \text{M}\Omega$$

Transconductance: G_{m2}

Suppose $G_{v16} \simeq 1 \rightarrow G_{m2}$ is then transconductance g_{m17} of Q_{17} with the 100 Ω (R_E) in the emitter:

$$G_{m2} = \frac{g_{m17}}{1+(g_{m17})\ R_E} = \frac{1}{147\Omega} = 7\ \text{mS}$$

1.5.5. *The third stage*

1.5.5.1. *Input resistance:* R_{e3}

The output resistance of the stage, R_{o2}, is that of Q_{13B} in // with that seen from the collector of Q_{17}:

$$R_{o2} = r_{o13B} // r_{o17} \cdot \left(\frac{1 + g_{m17} R_E}{1 + \frac{g_{m17} R_E}{\beta_o}} \right) \quad [1.38]$$

Here:

$$\beta_o \gg 1 \;\rightarrow\; R_{o2} = r_{o13\beta} // \{r_{o17}(1 + g_{m17} R_E)\} \quad [1.39]$$

Suppose $\eta_{PNP} = 5 \times 10^{-4}$:

h $_{NPN} = 2 \times 10^{-4}$

Hence:

$$r_{o17} \cdot g_{m17} = \frac{1}{\eta_p} = \frac{1}{2.10^{-4}} = 5{,}000 \qquad (R_E = 100\ \Omega)$$

$$r_{o13B}\ g_{m13B} = \frac{1}{\eta_p} = \frac{1}{5 \times 10^{-4}} = 2{,}000$$

and:

$$r_{o17}\ (1 + g_{m17} R_E) = r_{o17} + \eta_p^{-1} R_E$$

$$= r_{o17} + 5{,}000 \times 100$$

$$= r_{o17} + 500\ \text{k}\Omega$$

$$R_{o2} = r_{o13B} // \{r_{o17} + 500\ \text{k}\Omega\}$$

$$g_{r_{17}} = \frac{\beta_o}{r_{\pi 17}} = \frac{250}{11.8 \times 10^3} = 2.1 \times 10^{-2} \Rightarrow r_{o17} = \frac{1}{\eta_p \times g_{m7}} = \frac{5{,}000}{2.1} \cdot 10^{+2}$$
$$= 250\ \text{k}\Omega$$

$$r_{o13B} = \frac{1}{\eta_n\ g_{m13B}} \left(= \frac{2{,}000}{\frac{\beta_o}{r_{\pi 13}}} \right) = \frac{2{,}000}{\frac{I_{C_{13B}}}{26\text{mV}}} \quad [1.40]$$

$$= \frac{2{,}000 \times 26 \times 10^3}{550 \times 10^{-6}} = \sim 95\ \text{k}\Omega$$

$$R_{o2} = 95\ \text{k}\Omega\ //\ (750\ \text{k}\Omega) \simeq 85\ \text{k}\Omega$$

(see Figure 1.30).

1.5.5.2. *Output stage*

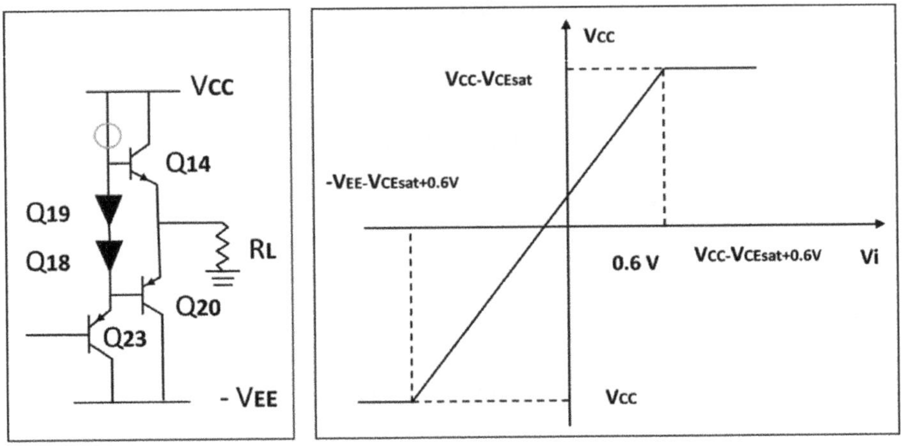

Figure 1.29. *Serial diode in emitter follower. For a color version of this figure, see www.iste.co.uk/gontrand/analog2.zip*

The output is either a source current or a sink current depending on the output voltage and charge.

In this case, the input resistance and the output resistance of this stage are highly dependent on the input currents and voltages. The input resistance of the output stage is much greater than the output resistance of the preceding stage, and therefore the value of the input resistance of the output stage hardly affects the voltage gain of the circuit.

Suppose, for example, that the output current is 2 mA and this current passes through a charge R_L (2 kΩ), Q_{14} conducts and Q_{20} provides a weak current.

Figure 1.29 (left) shows the output stage schematic; in this case, we are dealing with two emitter followers in series → $G_v \simeq 1$.

1.5.5.2.1. Input resistance R_{e3}

Let us first calculate the resistance seen from the base of Q_{14} called R_{eq2}.

Emitter follower:

$$\rightarrow \quad R_{eq2} = r_{\pi 14} + (\beta_{o14}+1)\, R_L \qquad [1.41]$$

$$= \left(\frac{I_{C14}}{26mV}\right)^{-1} \beta_{o14} + (\beta_{o14} + 1)R_L = \frac{26 \times 10^{-3}}{2 \times 10^{-3}}.250 + 251 \times 2\,\text{k}\Omega \sim 500\,\text{k}\Omega \quad [1.42]$$

Then, let us calculate the Thévenin equivalent resistance, R_{eq3}, seen from the emitter of Q_2 looking toward diodes Q_{19} and Q_{18}:

$$R_{eq3} = r_{d18} + r_{d19} + r_{o13A} \,//\, R_{eq2} \text{ (Figure 1.30)} \quad [1.43]$$

$$= 0 + 0 + 3 \times \frac{3.95\,\text{k}\Omega}{r_{o13}} \,//\, 500\,\text{k}\Omega \sim 180\,\text{k}\Omega$$

Thus, the input resistance of the stage is that of the emitter follower Q_{23} with a resistor R_{eq3} in its emitter:

$$R_{e3} = r_{\pi 23} + (\beta_{o23} + 1)\ R_{eq3} \quad [1.44]$$

$$= \frac{26\text{mV}}{157\mu\text{A}}.50 + (51).180\,\text{k}\Omega \sim 9\,\text{M}\Omega$$

– I_{c23} (current flowing in diodes) ~ 190 μA:

$$I_{c14} = 1\text{ mA}$$

If for an NPN, $\beta_o = 250$, and for a PNP, $\beta_o = 50$, then:

$$R_{e3} = r_{\pi 23} + 51 \times R_{eq3}$$

$$= \frac{\beta_{o,pnp}}{\frac{I_{C23}}{26\text{ mV}\ (v_t)}} + (51)\ R_{eq3} \quad [1.45]$$

$$= \frac{50}{187\times10^{-6}/26\ \times10^{-3}} + 51 \times 182\,\text{k}\Omega = 7\,\text{k}\Omega + 9{,}282\,\text{k}\Omega$$

$$R_{e3} \sim 9.3\ \Omega$$

The input resistance is much greater than the output resistance of the preceding stage; the amplifier is barely affected by the variations of the charge (of its resistance) connected to the output of the amplifier.

1.5.5.3. R_{S3} *output resistor*

The output resistance of the preceding stage must be introduced ($R_{o2} = 84\ \text{k}\Omega$).

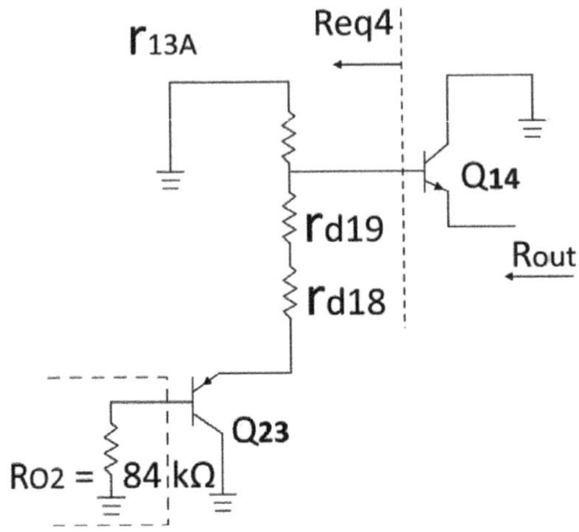

Figure 1.30. *Output resistance*

The resistance seen from the base of Q_{14}, looking to the left, is:

$$R_{eq4} = r_{o13A} // \left\{\left[r_{d19} + r_{d18} + \frac{R_{o2} + r_{\pi 23}}{\beta_{o23}+1}\right]\right\} \quad [1.46]$$

$$= 1.80k\ //\left(\frac{83k + 7K}{251}\right) = 20\Omega$$

The resistance seen from the output is then:

$$r_{out} = \frac{R_{eq4} + r_{\pi 14}}{\beta_{o14}+1} = 20\Omega\left(= \frac{1{,}8k + 3.25\ k}{251}\right) \quad [1.47]$$

We must add the current limiting series resistors:

$R_6 = 27\ \Omega.$

Hence:

$r_{out} = 20{+}27 = 47\ \Omega$

The output resistance changes with the rest point and is highly dependent on the current flowing in the output transistor.

1.5.5.4. *Performance small signals of the complete circuit*

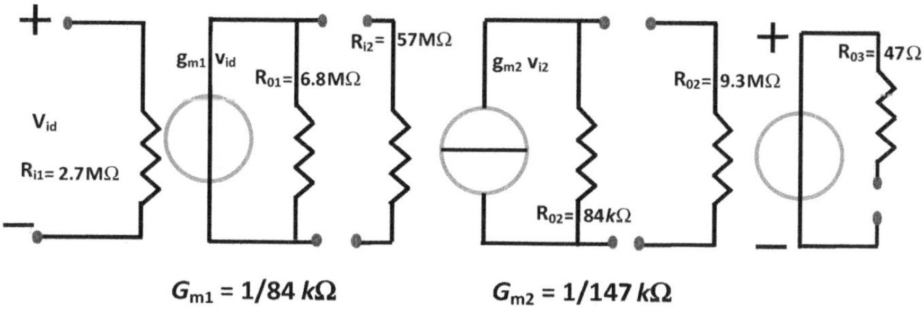

Figure 1.31. *The whole amplifier. For a color version of this figure, see www.iste.co.uk/gontrand/analog2.zip*

$$A_v = (G_{m1}).(R_{o1}//R_{i2}).\ (G_{m2}).(R_{o2}//R_{i3}) \quad [1.48]$$

$$= 574 \times 564 = 324{,}000.$$

The two stages have the same gain.

The stage charging the first reduces its gain by half. This leads to a dependence on β for the gain of the circuit; gain varies as a function of the temperature and the variations in the process.

However, the output stage does not significantly change the output of the second stage ($R_o << R_{i3}$) and the voltage gain is almost independent of the charge resistor output. The input and output resistances of the complete amplifier are:

$$R_{id} = 2.7\ \text{M}\Omega$$

$$R_o = 47\ \Omega$$

1.5.6. *Considering the effect of the second order: computer analysis*

The simplifying assumptions made above limit the accuracy of the results on the DC and the voltage gain.

1) Output resistances of the transistors are no longer being ignored.

The collector current typical of a lateral PNP increases by about 30% when the V_{CE} increases from 0 to 15 V, assuming that the emitter-base voltage is kept constant and $V_A = 50$ V.

For NPN devices, we have 12%, if $V_A = 120$ V.

The effect of this is to increase the bias currents in the circuit by 10%–30% compared to those calculated. This affects the calculated small-signal impedance levels.

The variation of β of transistors as a function of the current has been ignored. In fact, the current gain of a transistor drops for the low levels of collector currents due to the phenomenon of recombination in the space charge region of the emitter-base.

Because of this, the input devices Q_1 to Q_6 as well as Q_{16} have much lower current gains for small signals than assumed.

The consequence is reducing the voltage gain of the amplifier and increasing the input bias.

For example, SPICE 5: Simulator Program with Integrated Circuit Emphasis analog simulator (pioneering authors: Nagel, Vladimirescu), hence the typical corrections:

	I_c (mA)		I_c (mA)
Q_1	12.2	Q_{13A}	–222
Q_2	12.4	Q_{13B}	–683
Q_3	–12.3	Q_{14}	173
Q_4	–12.4	Q_{15}	Closed
Q_5	12.1	Q_{16}	17
Q_6	12.1	Q_{17}	688
Q_7	11.3	Q_{18}	–204
Q_8	–23.6	Q_{19}	16.4
Q_9	–30.6	Q_{20}	–173
Q_{10}	31	Q_{21}	Closed
Q_{11}	732	Q_{22}	Closed
Q_{12}	–701	Q_{23A}	–220
		Q_{23B}	Closed
		Q_{24}	Closed

1.6. Electric simulator

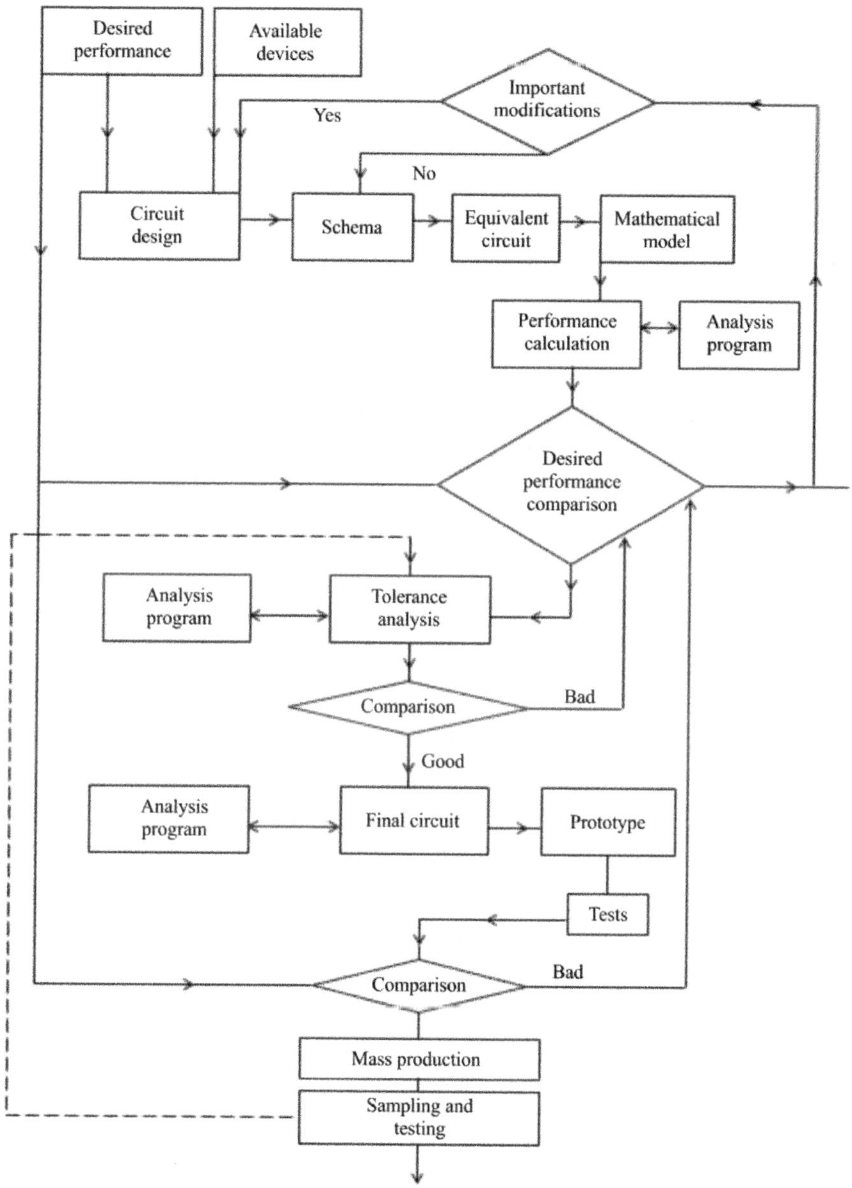

Figure 1.32. *Electrical simulator flowchart*

CAD programs for electronic circuits allow for a rapid calculation of the characteristics of a circuit that an electronics engineer has designed but has not made as a prototype.

These programs generally allow continuous, periodic or transient analysis. As a result, it is possible to provide for the operation of this circuit very quickly and, taking into account these results, to modify the device.

If a circuit has to be produced in a very large number of copies, it will be necessary to test its reliability and the evolution of its performance according to the tolerances on devices and environmental parameters.

Again, the speed of the computer calculation allows for a complete analysis "on paper" of the expected performance of the assembly.

The following flowchart gives a general procedure to be followed in the design of a circuit intended for mass production.

– Comments

The design of the analog circuit is not automatic. It is the electronics engineer who designs the circuit and the computer that analyzes it.

a) After the schematic diagram, an equivalent circuit is produced:

Example: equivalent circuit of a transistor from simple devices; the total circuit becomes more significant (more branches and nodes), but it is more easily analyzed on computer.

b) The constitution of the circuit is then given to the analysis program that supplies the performances of the circuit according to the chosen excitation mode.

c) Comparison of the performances with those desired directs the designer toward various modifications to be made.

– If the performances are similar, we move on to the analysis of tolerances.

– Sensitivity of the response is determined according to the evolution of the value of the device.

– Worst case operation:

- Circuit response based on device tolerances.

d) If, in theory, the operation is good, we can make a prototype, and test it.

If the test results are different from those expected, we return to the design of the schematic: effects may have been ignored (parasitic capacities, for example).

e) A test on a small series production will indicate whether the tolerance analysis has been valid.

1.6.1. *Analysis of steady-state linear circuits*

A circuit is formed by M branches and N nodes. From "input" data, we can calculate voltages with respect to a common point (usually ground) or currents. Basic relationships are KIRCHHOFF's laws.

The sum of the currents entering a node is equal to the sum of the outgoing currents. The sum of the voltages in a loop is equal to zero.

To determine the system of equations that describes a circuit, each element of it is first modeled in mathematical terms (ab initio, modeling: it is the ideas and the images that are born in our little brains).

This modeling step requires each element to be represented by an interconnection of branches of an ideal network. Each branch is then described by a mathematical relationship that defines the voltage or current in the branch as a function of variables unknown to the circuit.

If a circuit consists of M branches (1→M) and N nodes (0→N, the node "0" generally representing the ground), the matrix representation of Kirchhoff's law relating to the nodes is as follows:

$$[A][I_{b}] = [O] \quad [1.49]$$

Voltages of the branches in a network will be written as linear combinations of node voltages v_n with respect to the common point (the ground); we get:

$$[V_b] = [B] = [V_n] \quad [1.50]$$

Here, $[I_b]$ is the matrix of currents of general term $i\,j$ (current flowing in branch j) and $[A]$ is the incidence matrix:

$$[A] = \left[a_{k_j}\right] \; ; \; dim[A] = M \times N \, , a_{k_j} = \begin{cases} 1 \; if \; current \; i \, j \; leaves \; node \; k \\ -1 \; if \; current \; i \, j \; enters \; node \quad k \\ 0 \; if \; current \; i \, j \; does \; not \; pass \; through \; k \end{cases}$$

[A] is constructed from the circuit graph:

$$[B] = \left[b_{k_j}\right];\ b_{k_j} = \begin{cases} 1 \ if\ branch\ k\ leaves\ node\ \ j \\ -1 \ if\ branch\ k\ enters\ node\ j \\ 0 \ if\ branch\ k\ does\ not\ arrive\ at\ \ j \end{cases}$$

and we have: $\boxed{[B] = [A]^T}$

– Let us consider the example:

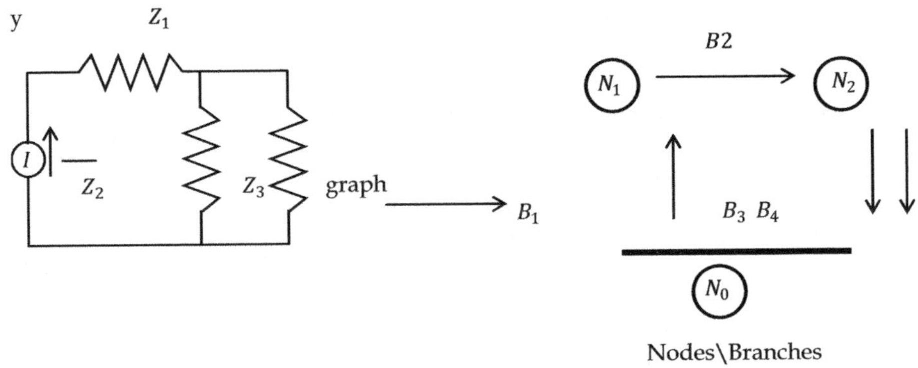

Figure 1.33. *From circuit to graph*

$$[\mathrm{A}] = \begin{bmatrix} -1 & 1\ 0\ 0\ 0 & -1 \\ & 1\ 1 & \end{bmatrix} \equiv \begin{matrix} N^1 \\ N^2 \end{matrix}\begin{bmatrix} B^1\ B^2\ B^3\ B^4 \end{bmatrix} \qquad [1.51]$$

Essentially:

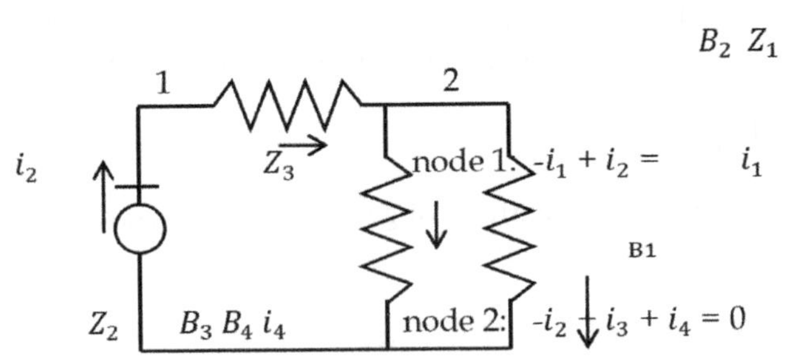

Figure 1.34. *Laws of nodes*

Or:

$$\begin{bmatrix} -1 & 1 & 0 & 0 & 0 & -1 \\ & & 1 & 1 & & \end{bmatrix} \begin{bmatrix} i^1 \\ i^2 \\ i^3 \\ i^4 \end{bmatrix} = [0] \quad [1.52]$$

Or:

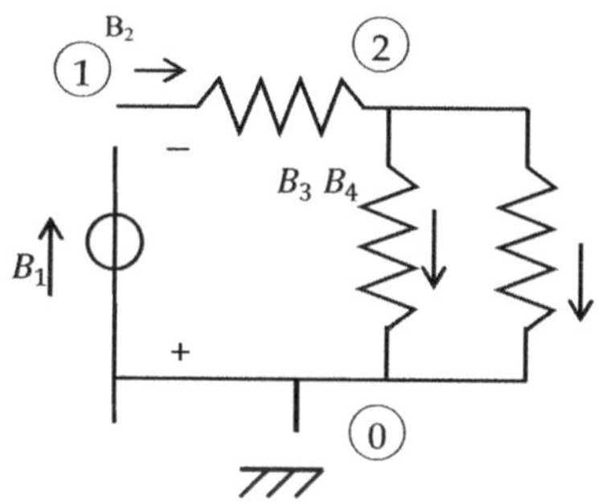

$$B_1 = v_{b_1} = v_0 - v_1 = -v_1$$

$$B_2 = v_{b_2} = v_1 - v_2 = v_1 - v_2 \quad [1.53]$$

$$B_3 = v_{b_3} = 2v0 - 2v3 = 3v3$$

$$B_4 = v_{b_4} = v_2 - v_0 = v_2$$

Figure 1.35. *Branch equations*

Or:

$$\begin{bmatrix} v_{b_1} \\ v_{b_2} \\ v_{b_3} \\ v_{b_4} \end{bmatrix} = \begin{bmatrix} -1 & 0 \\ +1 & -10 \\ & 10+ \\ & +1 \end{bmatrix} \begin{bmatrix} v_1 \\ v_2 \end{bmatrix} \quad [1.54]$$

In fact, there may be independent generators in the circuit (as in the example) (I_g):

$[I_B]$ becomes $[I_b] - [I_g]$

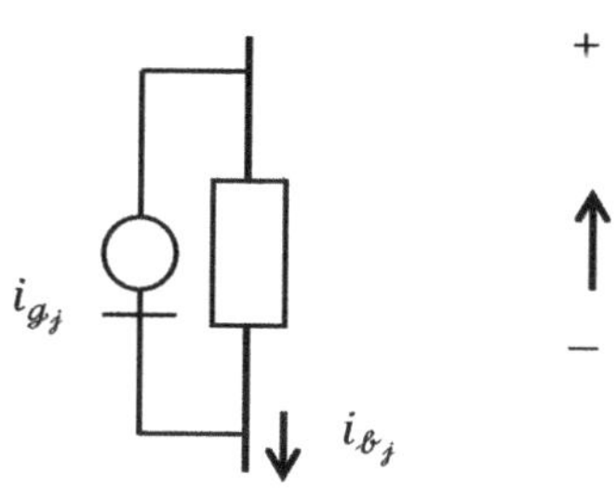

$$i^*_{b_j}[I_B] = [i^*_b] - [I_g] \quad [1.55]$$

Figure 1.36. *Current generator*

By introducing the admittance matrix, we get:

$$[I_b] = [Y_b]\,[V_b] - [I_g]$$

$$= [Y_b]\,[A]^T[V_n] - [I_g]$$

X^{ions} by $[A]$

$$[A]\,[I_b] = [A]\,[Y_b]\,[A]^T[V_n] - [A]\,[I_g]$$

$$= [0]$$

$$\boxed{[A]\,[Y_b]\,[A]^T[V_n] = [A]\,[I_g]} \qquad [1.56]$$

We write $[Y_n] = [A][\,Y_b][\,A]^T$ [1.56b]

where Y is the nodal admittance matrix.

In our case, we have:

$$[Y_b] = \begin{bmatrix} 0 & 0 & 0 & 0 \\ 0 & \frac{1}{Z_1} & 0 & 0 \\ 0 & 0 & \frac{1}{Z_2} & 0 \\ 0 & 0 & 0 & \frac{1}{Z_3} \end{bmatrix} \qquad [1.57]$$

and:

$$[I_g] = \begin{bmatrix} -I \\ 0 \\ 0 \\ 0 \end{bmatrix} \qquad [1.58]$$

The relationship giving the tensions of nodes is written as:

$$\begin{bmatrix} -1 & 1 & 0 & 0 \\ 0 & -1 & 1 & 1 \end{bmatrix} \begin{bmatrix} 0 & 0 & 0 & 0 \\ 0 & \frac{1}{Z_1} & 0 & 0 \\ 0 & 0 & \frac{1}{Z_2} & 0 \\ 0 & 0 & 0 & \frac{1}{Z_3} \end{bmatrix} \begin{bmatrix} -1 & 0 \\ 1 & -1 \\ 0 & 1 \\ 0 & 1 \end{bmatrix} \begin{bmatrix} v_{n_1} \\ v_{n_2} \end{bmatrix}$$

$$= \begin{bmatrix} -1 & 1 & 0 & 0 \\ 0 & -1 & 1 & 1 \end{bmatrix} \begin{bmatrix} -I \\ 0 \\ 0 \\ 0 \end{bmatrix} \quad [1.59]$$

$$[A] \; [Y_b] \; [A]^T \; [V_n] = [A] \; [I_g] \quad [1.60]$$

so:

$$\begin{bmatrix} \frac{1}{Z_1} & -\frac{1}{Z_1} \\ -\frac{1}{Z_1} & \frac{1}{Z_1} + \frac{1}{Z_2} + \frac{1}{Z_3} \end{bmatrix} \begin{bmatrix} v_{n_1} \\ v_{n_2} \end{bmatrix} = \begin{bmatrix} I \\ 0 \end{bmatrix} \quad [1.61]$$

Inversion $\Longrightarrow$

$$\begin{bmatrix} v_{n_1} \\ v_{n_2} \end{bmatrix} = \left[\frac{1}{Z_1} \times \left(\frac{1}{Z_2} + \frac{1}{Z_3}\right)\right]^{-1} \cdot \begin{bmatrix} \frac{1}{Z_1} + \frac{1}{Z_2} + \frac{1}{Z_3} & \frac{1}{Z_1} \\ \frac{1}{Z_1} & \frac{1}{Z_1} \end{bmatrix} \begin{bmatrix} I \\ 0 \end{bmatrix} \quad [1.62]$$

Of course, we find that:

$$v_1 = \left[Z_1 + \frac{1}{\frac{1}{Z_2} + \frac{1}{Z_3}}\right] I \text{ and } v_2 = \frac{I}{\frac{1}{Z_2} + \frac{1}{Z_3}} \quad [1.63]$$

It is of course possible to introduce independent voltage sources, as well as dependent sources, that is, current or voltage controlled current sources (SCC and SCV) and voltage or current controlled voltage sources (SVV and SVC); in these cases, branches may be dependent: $[Y_b]$, for example, is no longer a diagonal matrix.

In practice, we often deal with 50 or 100 equations, but software such as SPICE can solve several thousands; in fact, the matrices are "hollow" (sparse), that is, they contain a lot of zeros, and there are methods of reorganizing these matrices to invert them (solve them) without making unnecessary calculations.

Therefore, matrix-type resolution methods are used, whether the models are linear or linearized "around a point".

In the case of SPICE, algorithms are very "robust", that is to say they converge rather well; nevertheless, and this often because of a "mounting error" for example a lug of a transistor in the air (not referred to a node), the software will provide an erroneous result; in practice, this can be the last calculated iteration.

1.6.2. *Transitional analysis*

In a transient analysis, differential equations relating voltages and currents must be considered.

One method consists of solving the system of differential equations describing the network.

Another method consists of replacing the capacitors and inductors by an equivalent diagram valid for small variations in current and voltage for a time t.

The problem is then to solve, at each instant t_i, the linear equation system describing the circuit.

We return to a program of analysis in continuous quasi-static regime. We will first look at the latter case.

Models

Model for "choke"

The current–voltage relationship in a choke L is:

$$i(t) = \frac{1}{L}\int_0^t v(t)dt + i_0 \quad [1.64]$$

$$v(t)$$

$$\longleftrightarrow$$

$$v(t) = L\frac{di}{dt}$$

From the time t = 0, it is possible to calculate the value of $i(t)$ at instants: $t_0/t_1 - t_{R-1}, t_k$:

$$i(t) = i_k \text{ and } i_k = \frac{1}{L}\int_{t_{k-1}}^{t_k} v\,dt + i_{k-1} \qquad [1.65]$$

$$t_k - t_{k-1} = \Delta t, \text{small.}$$

By making a linear interpolation of $v(t)$ between t_k and t_{k-1}, we approximate the integral to the surface of a trapezium.

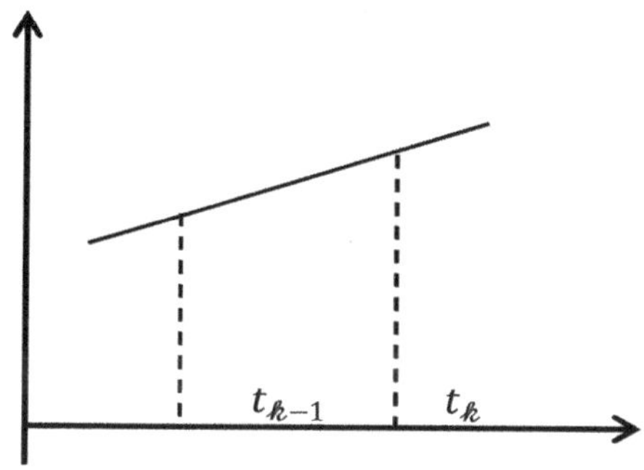

$$\int_{t_{k-1}}^{t_k} v\,dt = \frac{1}{2}\Delta t(v_k + v_{k-1})$$

and

$$i_k = \frac{1}{2L}\Delta t\,(v_k + v_{k-1}) \qquad [1.66]$$

Figure 1.37. *Trapeze surface approximation*

$$i_k = \left(i_{k-1} + \frac{\Delta t}{2L}\,v_{k-1}\right) + \frac{\Delta t}{2L}\,v_k \qquad [1.67]$$

We can write this expression in the form:

at time t_k: $\boxed{i = I + gv}$ [1.61], with $\begin{cases} g = \dfrac{\Delta t}{2L} \\ I = i_{k-1} + g.v_{k-1} \end{cases}$

Hence, the equivalent scheme:

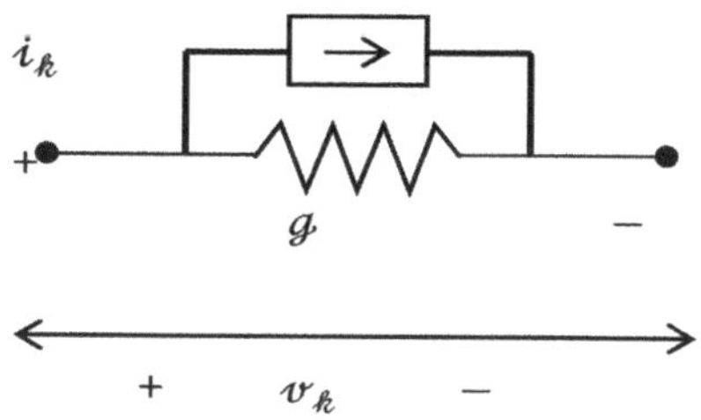

Note:

a) The direction of currents and voltages

b) The values of i at t_k are used to calculate the values of i and v at t_{k-1} → I

c) v_k is the voltage at the terminals of the choke here

Figure 1.38. *An equivalent scheme: a current source*

Capacitor:

$$i(t) = C\frac{dv(t)}{dt} \qquad [1.68]$$

and:

$$i_k = C\left(\frac{dv}{dt}\right)_{t=t_k}$$

Using the same approximation as in the case of the choke, we get:

$$i_k = \frac{C}{\Delta t}(v_k - v_{k-1}) \qquad [1.69]$$

$$i_k = g\,v_k - g\,v_{k-1} \qquad \text{where} \quad g = \frac{C}{\Delta t}$$

Equivalent diagram:

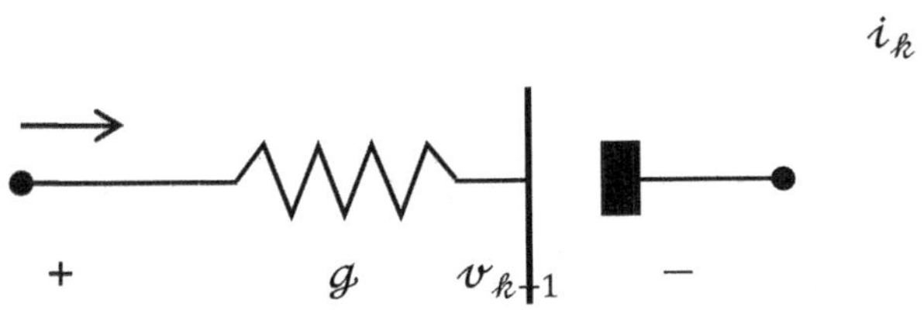

Figure 1.39. *Equivalent diagram*

In the capacitance and inductance model, we had developed v_k to the order of 1:

$$v_k = v_{k-1} + \Delta t\left.\frac{dv}{dt}\right|_{k-1}$$ [1.64], so linearize v around k.

In fact, for very good accuracy, it is necessary to have Δt "very" low. It is possible to increase the order of development if there is a problem of precision, but Δt must in any case remain low, hence a very long calculation time.

In practice, an implicit (and not explicit) diagram is currently used, the most widely used being the second-order Gear scheme (see SPICE):

$$v_k = \frac{4}{3}v_{k-1} - \frac{1}{3}v_{k-2} + 2\frac{\Delta t}{3}\left.\frac{dv}{dt}\right|_k \qquad [1.70]$$

Gear algorithm

The methods proposed by Gear are defined by the equation:

$$(1) \quad x_{n+1} = \sum_{i=0}^{k} \propto i \; x_{n+1-i} \quad : k + 1 \text{ coefficients} \qquad [1.71]$$

Calculation of coefficients

Suppose there is an interpolation polynomial:

$$y(t) = a_0 + a_1(t_{n+1} - t) \ldots\ldots + a_k (t_{n+1} - t)^k \qquad [1.72]$$

Coefficients a_i are obtained by smoothing [1.71] for the $k + 1$

Values $(x_{n+1}, x_n, \ldots\ldots, x_{n+1-k})$ are given as :

$$\begin{bmatrix} x_{n+1} \\ x_n \\ \\ \\ \\ x_{n+1-k} \end{bmatrix} = \begin{bmatrix} 1\,0\,0 & & & \ldots 0 \\ 1\; h\; h^2 & & & \ldots hk \\ . & . & . & . \\ . & . & . & . \\ . & . & . & . \\ 1\; kh\; (kh)^2 & & & \ldots (kh)^k \end{bmatrix} \begin{bmatrix} a_0 \\ a_1 \\ . \\ . \\ . \\ a_k \end{bmatrix} \qquad [1.73]$$

Substituting [1.72] in [1.70], we get:

$$-a_1 = \propto_0 \left[a_0 + a_1\, h + \ldots\ldots\ldots + a_k\, h^k\right]$$

$$+ \propto_1 \left[a_0 + a_1(2h) + \ldots\ldots\ldots + a_k\,(2h)^k\right] \qquad [1.74]$$

.

.

.

$$+ \propto_k \left[a_0 + a_1(kh) + \ldots\ldots + a_k\,(kh)^k\right]$$

Coefficients of the second part of [1.73] are equal to those on the left:

$$\begin{bmatrix} 1 & 1 & \ldots & 1 \\ h & 2h & \ldots & kh \\ \\ \\ h^k & (2h)^k & \ldots & (kh)^k \end{bmatrix} \begin{bmatrix} \propto_0 \\ \propto_1 \\ . \\ . \\ . \\ \propto_k \end{bmatrix} = \begin{bmatrix} 0 \\ -1 \\ . \\ . \\ . \\ 0 \end{bmatrix} \qquad [1.75]$$

The solution of [1.74] for the values of h from 1 to 6 are stepwise over the constant time:

Gear 1:

$$x_{n+1} = x_n + h\,\dot{x}_{n+1} - \frac{h^2}{2}\frac{d^2x}{dt^2} \qquad [1.76]$$

(Euler's formula)

Gear 2:

$$x_{n+1} = \frac{3}{4}\,x_n - \frac{1}{3}\,x_{n-1} + \frac{2h}{3}\,\dot{x}_{n+1} - \frac{2h^3}{9}\frac{d^3x}{dt^3} \qquad [1.77]$$

Gear 3:

$$x_{n+1} = \frac{18}{11}\,x_n - \frac{9}{11}\,x_{n-1} + \frac{2}{11}\,x_{n-2} + \frac{6h}{11}\,x_{n-3} - \frac{3h^4}{22}\frac{d^4x}{dt^4} \qquad [1.78]$$

Gear 4:

$$x_{n+1} = \frac{48}{25}\,x_n - \frac{36}{25}\,x_{n-1} + \frac{16}{25}\,x_{n-2} - \frac{3}{25}\,x_{n-3} + \frac{12h}{25}\,\dot{x}_{n+1} - \frac{12h^5}{125}\frac{d^5x}{dt^5} \qquad [1.79]$$

Gear 5:

$$x_{n+1} = \frac{300}{137}\,x_n - \frac{300}{137}\,x_{n-1} + \frac{200}{137}\,x_{n-2} - \frac{75}{137}\,x_{n-3} + \frac{12}{137}\,x_{n-4}\,\frac{60h}{137}$$

$$\dot{x}_{n+1} - \frac{10h^6}{137}\frac{d^6x}{dt^6} \qquad [1.80]$$

Gear 6:

$$x_{n+1} =$$
$$\frac{360}{147}x_n - \frac{450}{147}\,x_{n-1} + \frac{400}{147}\,x_{n-2} - \frac{225}{147}\,x_{n-3} + \frac{72}{147}\,x_{n-4} - \frac{10}{147}\,x_{n-5} +$$

$$\frac{60h}{197}\,\dot{x}_{n+1} - \frac{180h^7}{3087}\frac{d^7x}{dt^7} \qquad [1.81]$$

NOTE.– Orders 1 and 2 are stable: if the system is nonlinear, one of the most widely used methods is the Newton–Raphson method.

1.6.3. *Nonlinear system: Newton–Raphson method*

Let us solve the system of two nonlinear equations with two unknowns.

$$f(x,y) = 0 \quad \text{and} \quad y(x,y) = 0$$

We start from a test solution (choose as wisely as possible: (x_0, y_0)

We then write: $x = x_0 + \Delta x$ and $y = y_0 + \Delta y$

and:

$$f(x,y) = f(x_0,y_0) + \Delta x\, f'_{x_0} + \Delta y\, f'_{y_0} = 0$$

$$g(x,y) = g(x_0,y_0) + \Delta x\, g'_{x_0} + \Delta y\, g'_{y_0} = 0$$

It is a linear system with respect to Δx and Δy:

We write $f_0 = f(x_0,y_0)$ and $f'_{x_0} = f'_x(x_0,y_0)$ (same for g)

$$\Delta x = (g_0\, f'_{y_0} - f_0\, g'_{y_0}) / (f'_{x_0}\, g'_{y_0} - g'_{x_0}\, f'_{y_0})$$

$$\Delta y = (f_0\, g'_{x_0} - g_0\, f'_{x_0}) / (f'_{x_0}\, g'_{y_0} - g'_{x_0}\, f'_{y_0})$$

This provides $x_1 = x_0 + \Delta x$

$$y_1 = y_0 + \Delta y$$

The process is repeated until convergence.

This method can be generalized to a system of n equations with n unknowns.

For $G(X) = 0$ to be resolved (for the transient regime), we have to solve $G(X, \dot{X}, t) = 0$

That is:

$$\begin{cases} g_1\,(x_1, \ldots\ldots\, x_4) = 0 \\ \\ g_i\,(x_1, \ldots\ldots\, x_4) = 0 \\ g_r\,(x_1, \ldots\ldots\, x_4) = 0 \end{cases}$$

We start from a solution vector X_0 as close as possible to the guess solution:

$$G(X) = G(X_0 + \Delta X) = G(X_0) + [J(X_0)\Delta X] = 0$$

$[J(X_0)]$ is the Jacobian matrix J $[J(X)]$ of the system in $X = X_0$, that is:

$$[J(X)] = \begin{bmatrix} \frac{jg_1}{jx_1} & \frac{jg_2}{jx_2} & \dots\dots\dots & \frac{jg_1}{jx_n} \\ \\ \frac{jg_n}{jx_1} & \frac{jg_n}{jx_2} & \dots\dots\dots & \frac{jg_n}{jx_n} \end{bmatrix} \quad [1.82]$$

$$X = \{x_1, \dots\dots x_4\}$$

$$\Delta X = -[J(X_0)]^{-1} G(X_0)$$

The Jacobian inversion is typically performed by a Gear algorithm.

Secant method

In Newton–Raphson method, the Jacobian derivatives are approximated by finite differences:

$$[J(X)] = \begin{bmatrix} \frac{g_1(x_1) - g_1(x_0)}{x_1 - x_0} & \dots\dots\dots & \frac{g_1(x_n) - g_1(x_n)_{-1}}{x_n - x_0} \\ \\ \frac{g_n(x_1) - g_n(x_0)}{x_1 - x_0} & \dots\dots\dots & \frac{g_1(x_n) - g_n(x_{n-1})}{x_n - x_{n-1}} \end{bmatrix} \quad [1.83]$$

Simulation programs preferentially use the Newton–Raphson method rather than that of the secant.

Calculating derivatives by finite differences barely provides an advantage over their explicit calculations, while the Newton–Raphson algorithm converges faster than that of the secant, the Jacobian of which is an approximation of the latter (Newton–Raphson).

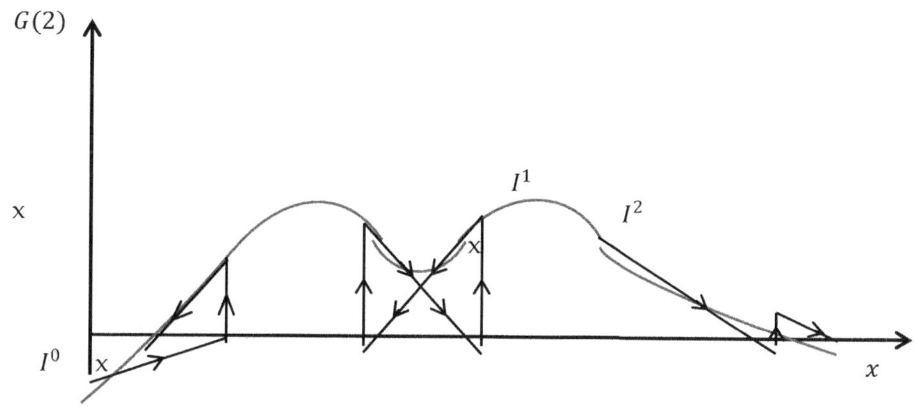

Figure 1.40. *Convergence to the solution. For a color version of this figure, see www.iste.co.uk/gontrand/analog2.zip*

Circuit performance analysis

A first solution to the problem is to study the sensitivity of the circuit to the variation in the value of a device.

The device or devices which play a predominant role in the response of a circuit and will determine the direction of variation of this response can then be determined.

A second method consists of simulating a random variation of the values of devices around their nominal value.

After a large number of simulations, it will be possible to construct a histogram that will describe the probable operation of the circuit. This method (also very common in simulation) is known as the Monte Carlo method.

Concept of Monte Carlo methods

These methods are used to simulate random phenomena and in numerical calculation to calculate integrals.

They are based on the generation of random or pseudo-random numbers. The latter are used in CAD; they are different from random numbers in that, when the first number or numbers are arbitrarily chosen, the whole sequence is completely finished: the sequence is then reproducible, which makes it possible to check the programs in which they are used.

The important point is that their properties are identical to that of random numbers. This, in fact, cannot be proven, but we can subject pseudo-random numbers to the same tests as random numbers, and use them if the tests are satisfied.

Calculation of π using this method

This is the famous Buffon's needle problem.

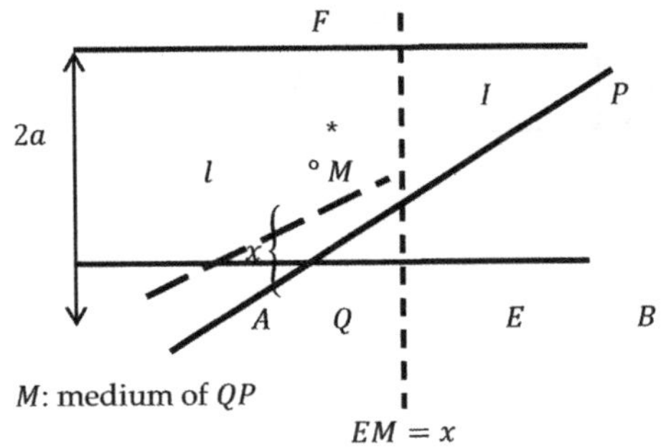

Let us consider either a sheet of paper with parallel straight lines separated by $2a$. A needle of length $2l$ is available which is dropped at random on the sheet $l < a$. We are going to calculate the probability that the needle will meet a parallel.

Figure 1.41. *Buffon's needle*

For M to cut AB, it is necessary that: $x < l$

The probability that M on EI is between x and $x + dx$ is:

$$P_1 = \frac{dx}{a}$$

In addition, it is necessary to have $\widehat{EMQ} < \widehat{EMQ'}$ $(MQ' = l)$:

$$\widehat{EMQ'} = arcos\frac{x}{l}$$

The probability that $\widehat{EMQ} < \widehat{EMQ'}$ is given as:

$$P_2 = \frac{2}{\pi}\ arcos\frac{x}{l}$$

$\frac{2}{\pi}$: normalization coefficient because $0 < EMQ' < \frac{\pi}{2}$

P_1 and P_2 are independent probabilities.

The total probability is:

$$P = \int_0^l \frac{2}{\pi} \quad arcos\frac{x}{l}\frac{dx}{a} \qquad [1.84]$$

$$(\pi y = \int_0^a = \int_0^l + \int_l^a \; ; \int_0^a = 0 \text{ because } l < a)$$

Let us write:

$$x = ly \qquad \Rightarrow y = \frac{x}{l}$$

$$\rightarrow P = \int_0^1 \frac{2l}{\pi a} \quad arcos\, y\, dy = \frac{2l}{\pi a} \qquad [1.85]$$

If we take $2l = a$, then:

$$\boxed{P = \frac{1}{\pi}} \qquad [1.86]$$

With 5,000 experiments, Ulli Wolff found that $\pi = 3.1596$. We can simulate this experiment provided we know how to draw numbers at random.

We present below an algorithm for generating a sequence of pseudo-random numbers.

One method, thanks to Lehmer, consists of using a sequence of numbers whose periodicity is great with respect to the quantity of numbers to be used (for a 32-bit computer, one could exceed 2^{32}, and thus create an artificial cutoff frequency).

A uniform distribution is obtained:

$$x_i = a\, x_{i-1} + c \quad \text{(modulo } m\text{)}$$

This means that number x_i is equal to the remainder of the division by m of $a\, x_{i-1} + c$, where m, a and c are constants.

a, x_i and c are comprised between *0* and m.

Additionally:

1) c and m are prime between them.

2) $a \equiv 1 \bmod (p)$ for each prime factor p of m.

For example, $m = 16$, $a = 3$, $c = 1$, $x_0 = 2$.

$a\,x_0 + c = 7$, divided by 16, quotient 0 remains $7 \rightarrow x_1 = 7$.

$a\,x_1 + c = 22$, divided by 16, quotient 1 remains $6 \rightarrow x_2 = 6$.

Here, we have 2, 7, 6, 3, 10, 15, 14, 11, 2, 7, 6, and the sequence is periodic.

Commonly, $c = 0$ and $m = 2^{M}$ (because a calculator works easily in binary) with $M \simeq 30 \rightarrow 40$; hence, a period of 2^{N-2} if a is an odd power of 5 (high) and x_0 is odd; $\rightarrow$ $x_i = a\,x_{i-1}$ (mod $m = 2^N$).

In practice, either a large computer with a machine word with a length of 32 bits. If the number exceeds 2^{31}, the left bit changes to 1, and the number is considered to be negative.

The modulo 2^{31} calculations are carried out; the sequence of numbers is brought back between 0 and 1.

Simulation of the performance of a circuit by a Monte Carlo method

The precision of devices, the control coefficients of sources and possibly the independent sources are taken into account. It will be assumed that the distribution of values is uniform in the range of variation thereof. In reality, the distribution around the nominal value is close to a Gaussian (normal distribution). To take this into account, it suffices to introduce a weighting function. It is also assumed that variations of the various parameters are independent (no correlation). The deviation of the value of a device or parameter from its nominal value is determined by means of a random number. Thus, we will have a random distribution of values with respect to nominal values:

For example:

– Case of a resistor whose precision is 100% = TOL

$$\frac{\Delta R}{R} = TOL$$

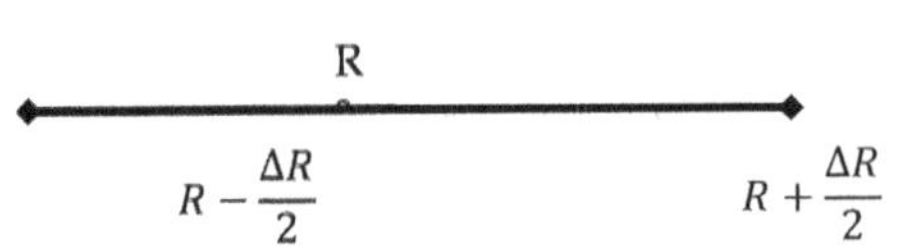

$R \rightarrow R + (0.5 - H)\,\Delta R$

$R \rightarrow R\,[1 + (0.5 - H)\,TOL]$

Where H is a random number between 0 and 1.

Figure 1.42. *Random distribution of R*

– Example of a simulation on a transistor mounted in E.C:

$$R \text{ at } 10\%,\ \ C \text{ at } 20\%\ \beta \text{ at } 30\%\ f = 1{,}000 \text{ Hg}$$

For a histogram of the output voltage see Figure 1.43, below:

Number of analyses: 20

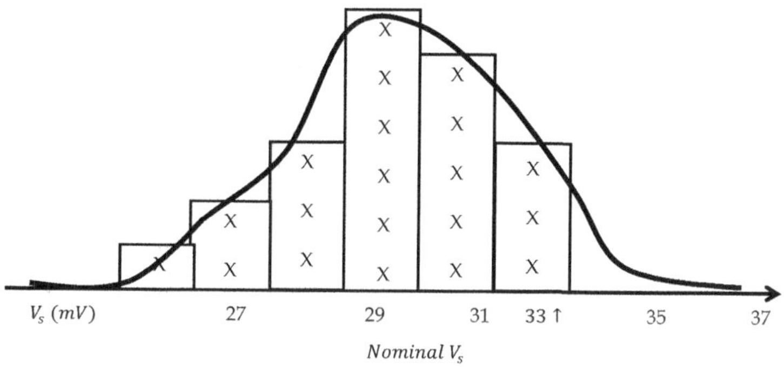

Figure 1.43. *Bell curve*

This histogram can be calculated automatically.

By means of a regression program, it is possible to determine the Gaussian describing a normal distribution:

$$f(V_s) = \frac{1}{\sqrt{2\pi t}} \quad e^{-\frac{(V_S - V_{SM})^2}{2\,t^2}} \qquad [1.87]$$

where $f(V_s)$: probability of having V_s and t: standard deviation.

By taking into account a weighting function in order to have a "normal" distribution in the range of variation of devices, it is possible to describe the probable performances of the circuit with fairly good precision. One of the drawbacks of this method is that it is necessary to repeat the analysis a large number of times, hence a fairly long calculation time.

Sensitivities of a circuit

The random simulation of the operation of the circuit can lead to fairly long calculation times. It is often sufficient to know the variation of a circuit voltage for a small variation (1% for example) of various devices of this circuit.

Let us consider parameters p (e.g. value of a resistance).

Node voltage: for a variation dp of the parameter p, we have:

$$[V_n](p_0 + dp) = [V_0] + [\Delta V_n] \quad [1.88]$$

$$[V_n](p_0 + dp) = [V_n(p_0)] + \sum_{i=1}^{K} \frac{J[V]_n}{Jp_i}\, dp_i \quad [1.89]$$

↑

summation on all parameters p are likely to vary.

where $V_n(p_0)$: values of voltages for the nominal values of parameters p.

For small variations around the nominal values of p, we limit ourselves to order 1 development.

This approximation is generally sufficient except for circuits where the variations of V_n are fast: very selective circuits, for example.

The quantity to be calculated is:

$$\boxed{[AV_n] = \sum_{i=1}^{K} \frac{J[V]_n}{Jp_i}\, dp_i} \quad [1.90]$$

To calculate the quantity $[AV_n]$, we will limit ourselves to circuits operating in continuous mode and consisting of G, R, I, E, and related sources β and Y_m.

We have the equation:

$$[Y_n]\,[V_n] = [A]\,\{I_g - [Y_b]\,[V_g]\}$$

or:

$$[V_n] = [Y_n]^{-1}[A]\{[I_g] - [Y_b][V_g]\} \quad [1.91]$$

a) Partial derivatives with respect to independent sources:

$$\frac{J[V_n]}{JI_{gj}} = [Y_n]^{-1}\,[A]\,\frac{J[I_g]}{JI_{gj}} = [Y_n]^{-1}\,[A][J_j] \quad [1.92]$$

where $[J_j] = \begin{cases} \text{Vector with 1 at line } j; \\ \text{0 elsewhere} \end{cases}$

$$\text{If: } \begin{bmatrix} 0 \\ 0 \\ 1 \\ 0 \\ 0 \\ 0 \end{bmatrix} \leftarrow i$$

b) Partial derivatives with respect to passive devices and related sources.

Here, we must consider variations of the matrix $[Y_{b.}]$:

$$[A][\,Y_b][\,A^t]\frac{J[V_n]}{Jp_j} + [A]\frac{J[Y_b]}{Jp_i}[\,A]^t[\,V_n] = -[A]\frac{J[Y_b]}{Jp_j}[\,V_g]$$

$$\frac{J[N_n]}{Jp_j} = -\{[A][\,Y_b][\,A^t]\}^{-1}[\,A]\frac{J[Y_b]}{Jp_j}\left\{[\,V_g] + \underbrace{[A^t][\,V_n]}_{[V_b]}\right\} \quad [1.93]$$

$$\boxed{\frac{J[V_n]}{Jp_j} = -\{[A][\,Y_b][A]^t\}^{-1}[\,A]\frac{J[Y_b]}{Jp_j}\{[\,V_g] + [V_b]\}} \quad [1.94]$$

We must therefore calculate the value of $\frac{J[Y_b]}{Jp_j}$.

Case of a resistor R_j:

$$\rightarrow \frac{J[Y_b]}{JR_i}$$

Matrix shape $[Y_b]$ with β: current source controlled by current:

$$[Y_b] = \begin{bmatrix} & i & \\ & \cdot & \\ & \cdot & \\ & {}^1\!/_{R_i} & \\ & & \\ & \beta_j\, i\ {}^1\!/_{R_i} & \end{bmatrix} \quad i$$

Recap: β_{ij} current source in branch j is controlled by current of branch i:

$$\frac{J[y_b]}{JR_0} = -\frac{1}{R_i^{\,2}}\,[j_i]\,[j_i]^t - \sum_{j\neq 1}\frac{\beta_{ji}}{R_i^{\,2}}\,[j_j]\,[j_i]^t \quad [1.95]$$

where:

– $[j_{i\,i}]$: hollow matrix with 1 in i_i;

– $[j_{j\,i}]$: hollow matrix with 1 in j_i.

Number of branches:

↓

$$\boxed{\frac{J[Y_b]}{JR_i} = -\frac{1}{R_i^2}\sum_{j=1}^{M} \beta_j\, i\, [j_j]\, [j_i]^t} \quad [1.96]$$

where $\beta_{i\,i} = 1$

and:

$$\boxed{\frac{J[V_n]}{JR_i} = [Y_n]^{-1}\, [A]\, \frac{1}{R_i^2}\sum_{j=1}^{M} \beta_j\, i\, [j_j]\, [j_i]^t \left\{[V_g] + [V_b]\right\}} \quad [1.97]$$

$$\boxed{\frac{J[V_n]}{JI_{g_i}} = [Y_n]^{-1}[A][j_i]}$$

$$\boxed{\frac{J[V_n]}{JV_{g_i}} = -[Y_n]^{-1}[A][Y_b][j_i]} \quad [1.98]$$

We will limit ourselves to sensitivities with respect to R_i, I_g, and V_g.

The method can be extrapolated to the other devices.

EXAMPLE.–

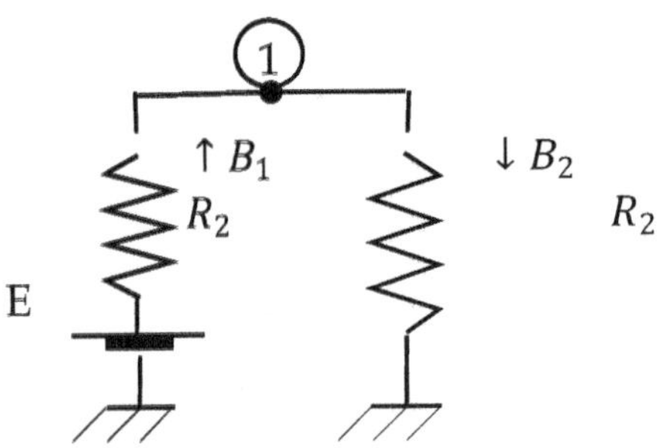

One node, two branches

$[A] = [-1 \;\; 1]$

$[V_g] = \begin{bmatrix} E \\ 0 \end{bmatrix} [Y_\beta] = \begin{bmatrix} 1/R & 0 \\ 0 & 1/R_2 \end{bmatrix}$

Figure 1.44. *A simple circuit. For a color version of this figure, see www.iste.co.uk/gontrand/analog2.zip*

$$[Y_\beta][A]^T \rightarrow \begin{bmatrix} 1/R & 0 \\ 0 & 1/R_2 \end{bmatrix}\begin{bmatrix} -1 \\ -1 \end{bmatrix} = \begin{bmatrix} -1/R \\ -1/R_2 \end{bmatrix} \quad [1.99]$$

$$[A][Y_b][A^T] = [-1 \;\; 1]\begin{bmatrix} -1/R_2 \\ 1/R_2 \end{bmatrix}$$

$$= [\,1/R_2 + 1/R_2] = [Y_n]$$

$$[Y_n] = \frac{R_1 + R_2}{R_1\,R_2} \Rightarrow Y_n{}^{-1} = \frac{R_1\,R_2}{R_1 + R_2}$$

$$[V_n] = [Y_n]^{-1}[A]\{[I_g] - [Y_b][V_g]\}$$

$$[Y_b][V_g] = \begin{bmatrix} 1/R_1 & 0 \\ 0 & 1/R_2 \end{bmatrix}\begin{bmatrix} E \\ 0 \end{bmatrix} = \begin{bmatrix} E/R_1 \\ 0 \end{bmatrix}$$

$$[V_n] = \begin{bmatrix} \frac{R_1\,R_2}{R_1 + R_2} \end{bmatrix} \times [-1 \;\; 1]\begin{bmatrix} -E/R_1 \\ 0 \end{bmatrix} = \frac{R_1\,R_2}{R_1 + R_2} * \;[E/R_1] \quad [1.100]$$

$$[V_n] = \frac{R_2\,E}{R_1 + R_2} = V_1$$

$$[V_b] = [A]^T[V_n] = \begin{bmatrix} -1 \\ 1 \end{bmatrix}\frac{R_2}{R_1 + R_2}\;E \quad [1.101]$$

$$\rightarrow [V_{b1}] = \frac{-R_2}{R_1 + R_2}\;E$$

$$[V_{b2}] = \frac{R_2}{R_1 + R_2}\;E$$

calculate $\dfrac{J[V_n]}{J\,R_1}$

$$\frac{J\,[V_n]}{J\,R_1} = \frac{R_1\,R_2}{R_1 + R_2}\,[-1 \;\; 1]\frac{1}{R_1{}^2}\begin{bmatrix} 1 & 0 \\ 0 & 0 \end{bmatrix}\begin{bmatrix} E \;-\; \frac{R_2}{R_1 + R_2}\;E \\ \frac{R_2}{R_1 + R_2}\;E \end{bmatrix}$$

$$= \frac{R_2}{R_1}\,\frac{1}{R_1 + R_2}\;[-1 \;\; 1]\;\begin{bmatrix} E \;-\; \frac{R_2}{R_1 + R_2}\;E \\ 0 \end{bmatrix}$$

$$= \frac{R_2}{R_1}\,\frac{1}{R_1 + R_2}\; - \;E\;\left[1 - \frac{R_2}{R_1 + R_2}\right] \quad [1.102]$$

$$= \frac{R_2}{R_1} \frac{1}{R_1 + R_2} - E - \frac{R_2}{R_1 + R_2}$$

$$\Rightarrow \frac{J\,[V_n]}{J\,R_1} = -\frac{ER_2}{(R_1 + R_2)^2}$$

In this simple case, it would have been possible to directly calculate:

$$\frac{J\,V}{J\,R_1} V_1 = E \frac{R_2}{R_1 + R_2} \Rightarrow \frac{J\,V_1}{J\,R_1} = -ER_2 \times \frac{1}{(R_1 + R_2)^2}$$

Let us say:

$$E = 10\text{ V} \qquad R_1 = 100\,\Omega \qquad \mathrm{R}_2 = 10\,\Omega$$

$$\frac{J[\,V_1]}{J\,R_1} = -10 \times \frac{10}{10^4} = -10^{-2}\ \mathrm{V}/\Omega$$

From a variation of the resistance $\Delta R_1 = 1\Omega$ results a variation ΔV_1 of $10^{-2}\,V$ with a sign opposite to ΔR.

We would then have:

$$[\Delta V_n] = \sum_{i=1}^{K} \frac{J[\,V_n]}{J\,P_i}\, dP_i$$

It can be written in the following form:

$$[\Delta V_n] = \sum_{i=1}^{K} \frac{J[\,V_n]}{J_i} \frac{P_i \Delta P_i}{100} \qquad [1.103]$$

where ΔP_i is the relative variation in % of the parameter P_i:

$$\boxed{[\Delta V_n] = \sum_{i=1}^{K} \left[S\,\frac{V_n}{P_i} \right] \Delta P_i} \qquad [1.104]$$

where $S\,\frac{V_n}{P_i}$ is the sensitivity of circuit node voltages to variations in the parameter P_i:

$$\left[S\,\frac{V_n}{P_i} \right] = \frac{J[\,V_n]}{J_{pi}} \frac{P_i}{100}$$

$[S]$ a column matrix.

Other definitions of sensitivity are possible.

In the preceding example, we have:

$$\left[S\frac{V_1}{R_1}\right] = -10^{-2} \times \frac{100\Omega}{100} = -[10^{-2}]$$

For:

$$\Delta R_1 = 1\% \qquad \text{that is,} \qquad \Delta R_1 = 1\Omega$$

$$\rightarrow \Delta V_1 = -10^{-2} V$$

Sensitivity analysis is generally possible with a large number of simulation programs; it is not always possible to specify the parameter with respect to which the sensitivity is desired.

Finding the worst case

This is an important point in the design of systems that only serve once (missiles, rockets) or equipment that can jeopardize the safety of people (public transport, for example).

We can describe this case of the Monte Carlo analysis, but we are not sure that we have reached the most malfunctioning, and the analysis can be long.

It is often preferable to use the information given by sensitivities and in particular by the values of partial derivatives, which we have seen in the previous paragraph.

Let us take the simple example seen earlier:

We have $\dfrac{J V_1}{J R_1} < 0 \quad and \quad \dfrac{J V_1}{J R_2} > 0 \quad for \ E > 0$

And suppose that R_1 and R_2 can vary between two limits.

From the respective signs of R_2 and R_1, it is evident that the minimum values of V_1 are:

$$V_{1m} = V_1 (R_{1M}, R_{2m})$$

$$V_{1M} = V_1 (R_{1m}, R_{2M})$$

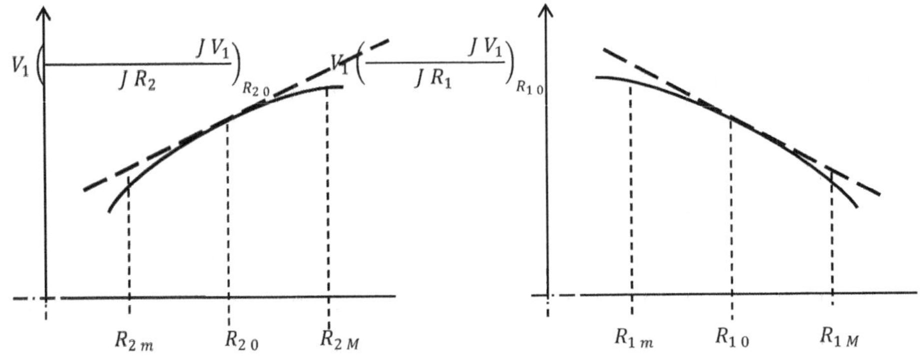

Figure 1.45. *Partial derivative values*

More generally, for a parameter p and a response f of a circuit, there is the most malfunctioning, that is, $(P)f_m$ and $(P)f_M$:

$$(P)f_m = \begin{cases} P_m \ if \dfrac{jf}{jp} > 0 \\ P_M \ if \dfrac{jf}{jp} < 0 \end{cases} \quad \text{and} \quad (P)f_M = \begin{cases} P_M \ if \dfrac{jf}{jp} > 0 \\ P_m \ if \dfrac{jf}{jp} < 0 \end{cases}$$

It is possible to calculate f_m and f_M from the slopes of the curves

Calculated at the first order:

$$f(P_M) = f_M = f_0 + \textstyle\sum_{i=1}^{P} \ \Delta f_i \qquad [1.105]$$

$$\uparrow$$

$$\text{f}(p0)$$

$$if \quad \Delta f_i = \begin{cases} \dfrac{jf}{jp_i}(p_{iM} - p_0) & if \dfrac{jf}{jp_i} > 0 \\ \dfrac{jf}{jp_i}(p_0 - p_{im}) & if \dfrac{jf}{jp_i} < 0 \end{cases}$$

Then, we have the greatest value.

Similarly, for the smallest value:

$$(f_M = f_0 + \sum_{i=1}^{P} (\Delta f_i))$$

$$(\Delta f_i) = \begin{cases} \frac{jf}{jp_i}(p_{iM} - p_0) & if \frac{jf}{jp_i} > 0 \\ \frac{jf}{jp_i}(p_0 - p_{im}) & if \frac{jf}{jp_i} < 0 \end{cases}$$

The relationships remain valid as long as the variation of f around the nominal value p_0 is not too fast; this is not the case for a circuit with too "sharp" a response, for example.

We considered the simple case where the value of $\frac{jf}{jp}$ did not vary in the domain $[P_m, P_M]$.

More complex cases may arise as shown diagrammatically in the following figures:

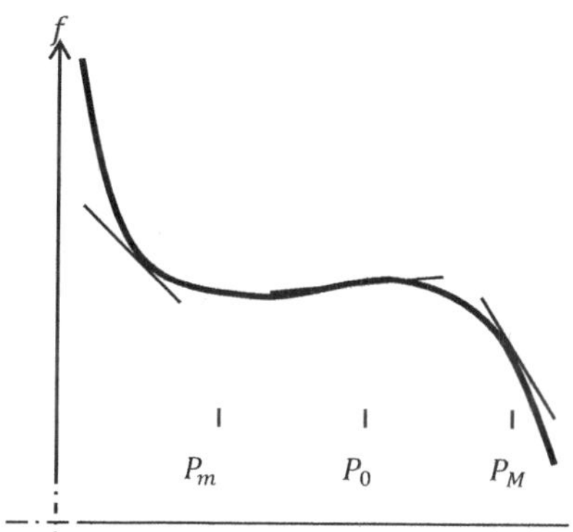

Changing the sign of the slopes $\left(\frac{jf}{jp}\right)_{p0}$: no longer valid for calculating f_m and/or f_M.

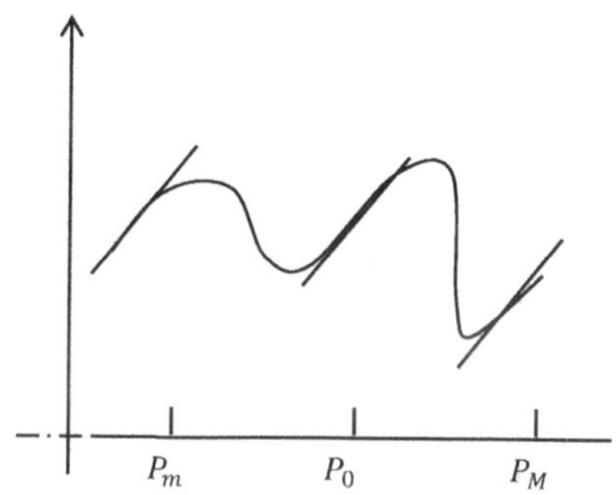

The slopes are the same but there is a change between these points.

Figure 1.46. *Atypical responses of a circuit*

In these cases, we interpolate $f(p)$.

SPICE Simulator

The software recognizes a number of devices for which models have been developed (see Volume 1). These models constitute a library which is used to constitute the input file.

Notations used for element names are as follows:

– R: resistance;

– C: capacitance;

– L: inductance;

– K: mutual inductance;

– T: delay line;

– V: independent voltage source;

– I: independent current source;

– G: voltage controlled current source (linear or otherwise);

– E: voltage controlled voltage source (linear or otherwise);

– F: current controlled current source (linear or otherwise);

– H: current controlled voltage source (linear or otherwise);

– D: diode;

– Q: bipolar junction transistor (BJT);

– J: JFET;

– M: MOSFET;

– X: Subcircuit;

Type of analysis

– Continuously nonlinear (DC);

– Nonlinear in transient (TRAN);

– Alternating linear (small signals) (AC);

– Fourier series and distortion (FOUR);

– Intermodulation (DISTO);

– Noise (NOISE);

– In temperature (TEMP);

– Continuous sensitivity (SENSE);

– Depending on parameters (VARY).

The actual electrical diagram is drawn. Each node is numbered in order to locate the access terminals to each device (the potential reference terminal always has the number 0).

Following this first step, we give the software all of the information necessary for the simulation. We thus constitute the input file that mainly contains the models used.

The simulation can then be started. The results are stored in an output file that we can read or process via a curve plot.

The advantage of the simulation thus appears in the speed of obtaining the frequency characteristics of the assembly for different values of elements.

We will now consider the variation of some technological parameters having an influence on the current or frequency gain characteristics of the device (the "process" and "device" people work together in companies).

a) Thickness of the base of the bipolar junction transistor.

The transition frequency is inversely proportional to the square of the thickness of the base [$f_t = \frac{1}{2\pi\tau}_{base}$ where:

$$\tau_{base} = \frac{1}{2}\frac{{W_B}^2}{D}$$

where W_B is the base width and D is the diffusivity of minority groups.

The geometrical parameter therefore acts directly on one of the data to be supplied to the simulator.

b) Doping of the collector of the bipolar junction transistor.

The capacitance between the collector and the base is that of the transition zone of the reverse-biased junction. The value, a function of the doping of the collector layer on which the ddp depends and of the surface of the device on which the current depends, is directly influenced by Miller effect on the frequency response of the circuit.

$$\left[C_{b_c} = \frac{C_{b_c}}{\left(1 - \frac{V}{\Psi_0}\right)^n} ; 0.33 < n < 0.5 ; \; C_{b_c} = \frac{A}{\Psi_0^n} \cdot \left[q \frac{\in}{2} \frac{N_A \, N_P}{(N_A + N_0)}\right]\right]$$

1.7. Simulation of a system with several active devices

Below, we will focus on the main functional analog blocks.

As a flagship example, we consider the conventional assembly below: a differential pair amplifier stage. Using the parameters of the NPN transistor which are given, the input file of the simulator is established. Continuous simulation makes it possible to determine the bias of each transistor (see quiescent or operational point), before carrying out an alternating analysis of small signals.

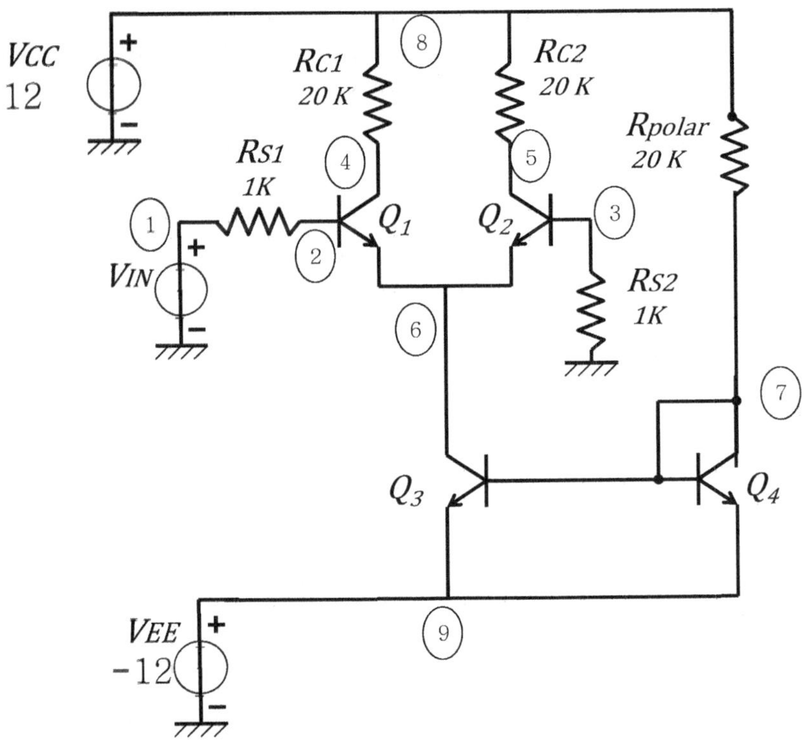

Figure 1.47. *Diagram of a differential pair*

The current source is then modeled by a current generator and its resistance is parallel (infinite for an ideal generator).

Corresponding SPICE program:

DIFFPAIR CKT _ SINGLE DIFFERENTIAL PAIR

*AC DEC 1 10 GHZ
*DC VIN _- 0.25 0.25 0.005
*TRAN DEC 1 10 GHZ

- (TRAN) DEC 1 10 GHZ
  VIN 1 0 1 0SIN(O 0.1 5 MEG 5NS) AC 1
  VCC 8 0 12
  VEE 9 0 -12
  Q1 4 2 6 QNL
  Q2 5 3 6 QNL
  RS1 1 2
  RS2 3 0
  RC1 8 4
  RC2 8 5
  Q3 6 7 9 QNL
  Q4 7 7 9
  Rpolat 7 8 20k
- OUTPUT V4 4 0 PRINT DC TRAN
- OUTPUT V5 5 0 PRINT MAG PHASE DC TRAN PLOT MAG PHASE DC TRAN
- MODEL QNL NPN (BF = 80 RB = 100 CCS= 2PF TF=0.3NS TR=0.6NS, CJE =3 PF,
  + CJC=2PF VA=50)
- END

Table 1.1. *Differentially mounted coupled emitters: SPICE program*

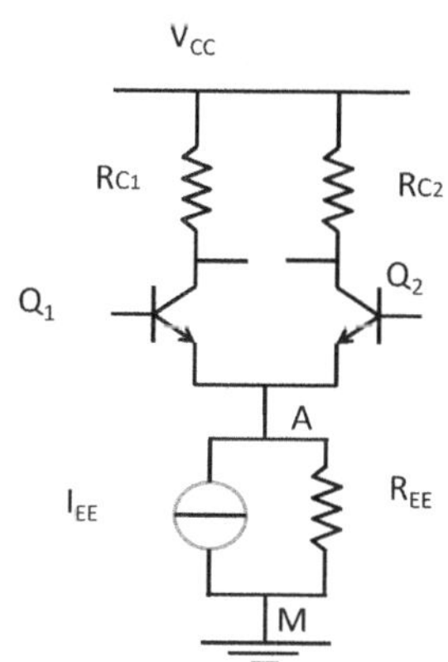

Figure 1.48. *"Differential" amplifier. For a color version of this figure, see www.iste.co.uk/gontrand/analog2.zip*

Important parameters are as follows:

– Rid;

– CMRR;

– differential gain (in mode): Adm.

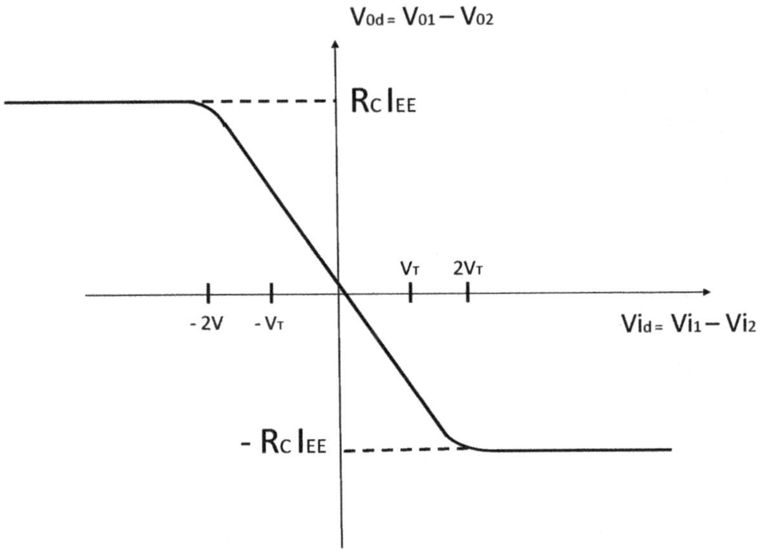

Figure 1.49. *Input/output characteristic*

Resistors increase the linearity range (see emitter feedback) but decrease the voltage gain (of the slope).

$I_{E1}=I_{E2}$, $I_{B1}=I_{B2}$, $I_{C1}=I_{C2}$,

$V_{B1E1} = V_{B2E2} = - V_{AM}$

$1VCE_{1= VCE2}$

$V_{B1E1}= V_{B2E2}$

Expression of V_{C2M}:

$V_{C2M} = V_{CC} - R_{C2}I_{E2}$;

$V_{C2M} = V_{CC} - R_{C1}I_{E2}$.

What is the current I' flowing through R_{EE}?

$$I' = V_{AM}/ R_{EE} = -VB_{E2}/ R_{EE}$$

In A: law of nodes:

$$I_{E1} + I_{E2} = I + I' = I - V_{B2E2}/R_{EE}$$

$$2I_{E2} = I - V_{B2E2}/ R_{EE} \rightarrow I_{E2} = I/2(I - V_{B2E2}/R_{EE})$$

Therefore:

$$V_{C2M} = V_{CC} - R_C I/2 + R_C V_{B2E2}/2 R_{EE}$$

$v_1^+ = v_1^- = 0$ (grounded transistor bases), but there is V_{C2M}.

When alternating voltages v_1^+ and v_1- are applied between bases and emitters, v_2 is not purely alternating but comprises a direct device V_{C2M}.

Studying the amplifier

$$r_1 = r_2 = r$$

$$b_1 = b_2 = b$$

In AC: Vdc is a short-circuit (the variations of a DC voltage are zero, or a battery can be likened to a large capacitance, of zero impedance if the latter is "infinite") and I is an open circuit.

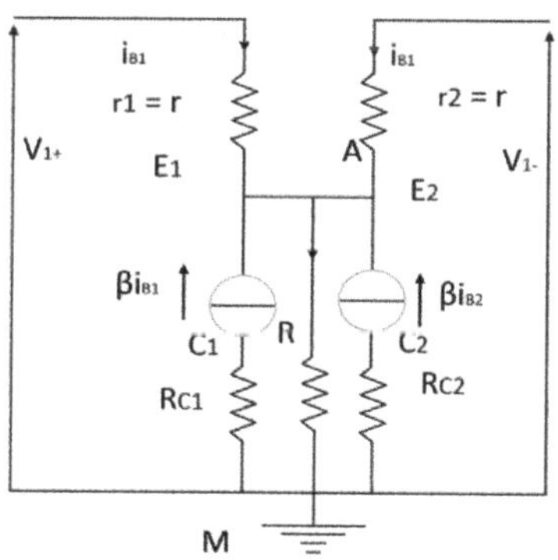

Figure 1.50. *Equivalent diagram. For a color version of this figure, see www.iste.co.uk/gontrand/analog2.zip*

Output voltage:

$$v_2 = -R_{C2}.\, i_{C2} = -\beta\, R_C\, i_{B2}$$

In A: law of nodes:

$$i = i\,\beta_1 + i\,\beta_{2+}\,\beta i\beta_{1+}\,\beta i\,\beta_{2+}\beta i\,\beta_2 = (\beta+1)i\,(\beta_{1+}\,i\,\beta_2)$$

$$v_1^{+} = r\,i\beta_2 + Ri = r\,i\beta_2 + R(\beta+2)(i\beta_1 + i\beta_2)$$

Therefore:

$$v_1^{-} = r\,i\beta_1 + Ri = r\,i\beta_1 + R(\beta+1)(i\beta_1 + i\beta_1)$$

$$v_1^{+} - v_1^{-} = r\,(i\beta_1 - i\beta_1)$$

$$v_1^{+} + v_1^{-} = r + R(\beta+1)(i\beta_1 + i\beta_1)$$

$$-\,i\beta_1 - i\beta_1^{\,-} = (v_1^{+} - v_1^{-})/R$$

and:

$$-\,i\beta_1 + i\beta_2^{\,-} = (v_1^{++}\,v_1^{-})/(r + 2R(\beta+1))$$

As a result:

$$i\beta_1 = (v_1^{+} - v_1^{-})/2r + (v_1^{+} + v_1^{-})/2\,(r + 2R(\beta+1))$$

Therefore:

$$v_2 = \ \beta Rc/(r + 2R(\beta + 1) - \beta Rc/(r + 2R(\beta+1)).\ (v_1^{+} + v_1^{-})/2$$

$$v_2 = \beta Rc/(v_1^{+} - v_1^{-})\,_{-}\,\beta Rc/(r + 2R(\beta+1)).\ (v_1^{+} + v_1^{-})/2 \qquad [1.106]$$

Cl:

V_2 depends on $v_1^{+} - v_1^{-}$, as well as on $v_1^{+} + v_1^{-}$

Performance of a difference amplifier

$A_d = \beta Rc/2r$ and $A_c = -\beta Rc/(r + 2R(\beta+1))$

$$v_1 = A_d.(v_1^{+} - v_1^{-}) + A_c.(\,v_1^{+} + v_1^{-})/2$$

Rejection rate

$$K = 20\log\left[\frac{A_d}{A_c}\right]$$

a.n:

$$A_d = 100,\ A_c = -0.2$$

Therefore, K = 60 dB

If v_1^+ and v_1^- are in phase and $v_1^+ - v_1^- = 1$ mV, then:

$v_1 = 0.198$ V ➔1% error.

If Ac is zero, $v_2 = 0.2$ V, then:

With $v_1^+ = 500.5$ mV and $v_1^- = 499.5$ mV:

$v_1 = 0.1$ V

Then: 50% error.

The output voltage no longer correctly reproduces the input voltage difference.

Input impedance

Differential mode

$Z_{ed} = v_d/i_d$

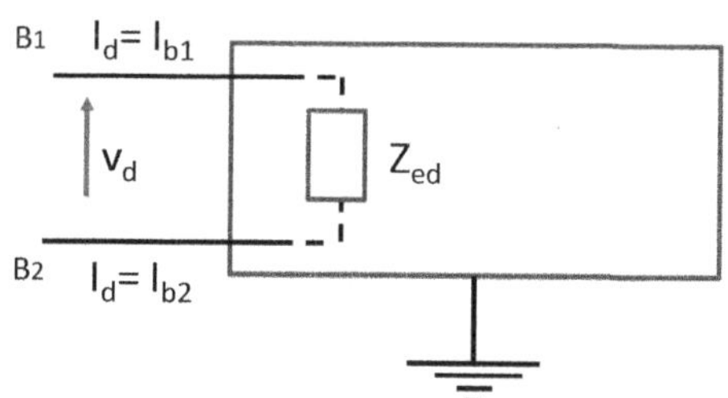

Figure 1.51. *Input impedance of a quadripole. For a color version of this figure, see www.iste.co.uk/gontrand/analog2.zip*

$v_d = v_{B1B2} = r.ib_1 - r.ib_2$

$i_d = i_{B1} - i_{B2}$ ➔ $V_d = 2r.\ i_d$

Input impedance

Common mode

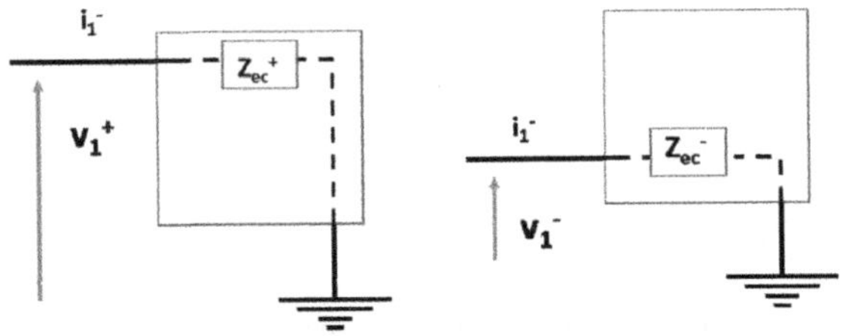

Figure 1.52. *Input impedances, relative to each input, in common mode. For a color version of this figure, see www.iste.co.uk/gontrand/analog2.zip*

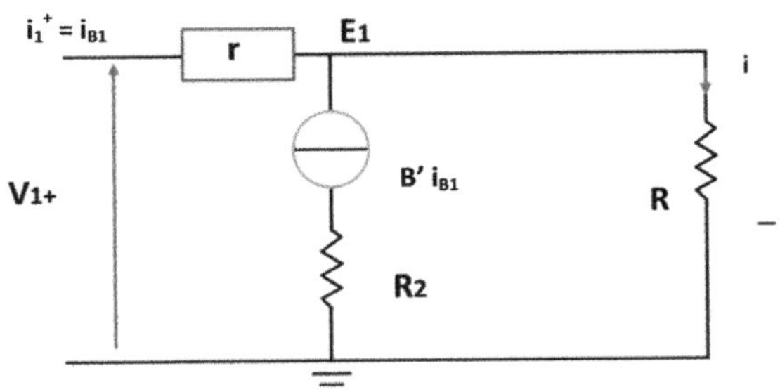

Figure 1.53. *Equivalent circuit. For a color version of this figure, see www.iste.co.uk/gontrand/analog2.zip*

$$Z^+_{ec} = v_1^+/i_1^+ \text{ and } Z^-_{ec} = v_1^- /i_1^-$$

$$Z^+_{ec} = Z^-_{ec}$$

Let us calculate R^+_{ec} by applying a single voltage between B and M.

$$v_1{+} = ri_{B1}{+}Ri$$

$$i_{B1} = i_1^+$$

$$i = (\beta{+}1)\, i_{B1}{=}(\beta{+}1)\, i_1^+$$

$$v_1{+} = r\, i_1^+\, R(\beta{+}1)\, i_1^+$$

$$R^+_{ec} = R^+_{ec}{=}\, r + R(\beta + 1)$$

Performance improvement

Ac = 0; it is then necessary to R → ∞.

Red = r and $R^{+}_{ec} = R^{-}_{ec} \rightarrow \infty$

Increase in R

A voltage generator can be connected in series with a high resistance.

If I = 1 mA R = 1 MΩ → RI = 1 V

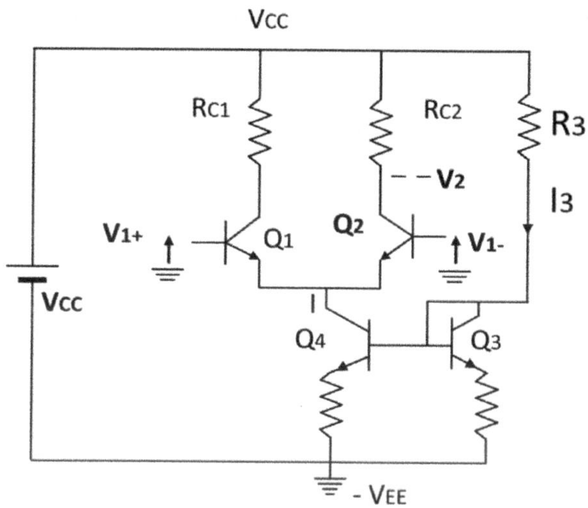

Figure 1.54. *A differential amplifier and its current source*

$V_{cc} = R_1 I_3 + V_{BE} + R_3 I_3$

$V_{BE3} + R_3 I_3 = V_{BE4} + R_4 I_4$

$I_3 = (V_{cc} - V_{BE3})/(R_1 + R_3)$

$I' = R_3/R_4 \cdot I_3 + (V_{BE4} - V_{BE3})/R_4$ [1.107]

I_4 is therefore constant: current generator.

R is several MΩ.

A_{DM} gain: differential mode

The active charge allows for a high gain:

$$A_{DM} = \left|\frac{V_0}{V_{DM}}\right| = \left|\frac{\beta_0 R_L}{r_\pi}\right| = g_m R_L$$

$v_{DM} = v_0/2$; output voltage in differential mode:

$$v_0 = (g_m R_L/2)\ v_{DM}$$

The transconductance is of the order of that of a differential stage equal to one-fourth of that of a bipolar junction transistor by a current I_0.

Bipolar junction transistors with coupled emitters

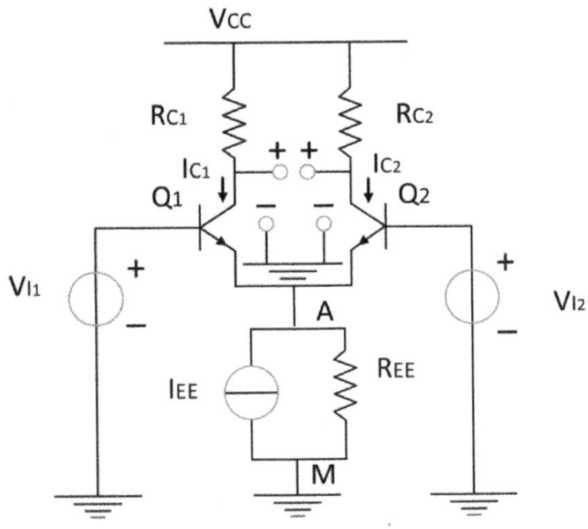

Figure 1.55. *Bipolar junction transistors with coupled emitters. For a color version of this figure, see www.iste.co.uk/gontrand/analog2.zip*

DC characteristic

The output resistance of the emitter current source R_{EE} is assumed to be infinite.

The base resistance is negligible and the output resistance of each transistor is infinite, not a very valid hypothesis, in particular, for small signals.

We have:

$$V_{i1} - V_{be1} + V_{be2} - V_{i2} = 0$$

Ebers–Moll model is used if v_{be1} and v_{be2} are much higher than V_T.

Then:

$$V_{be1} = V_T \log\frac{I_{c1}}{I_{c2}} \text{ and } V_{be2} = V_T \log\frac{I_{c2}}{I_{c1}}$$

If $I_{C1} = I_{C2}$:

$$\frac{I_{c1}}{I_{c2}} = \exp\left(\frac{V_{11-}\ V_{12}}{V_T}\right) = exp\left(\frac{V_{id}}{V_T}\right)$$

$$I_{c1}+I_{c2} = I_{EE} = \frac{1}{\alpha_F}(I_{c1}+I_{c2})$$

Therefore:

$$I_{c1} = \frac{\alpha_F I_{EE}}{1 + exp\left(\frac{-V_{id}}{V_T}\right)}$$

$$I_{c2} = \frac{\alpha_F I_{EE}}{1 + exp\left(\frac{+V_{id}}{V_T}\right)} \qquad [1.108]$$

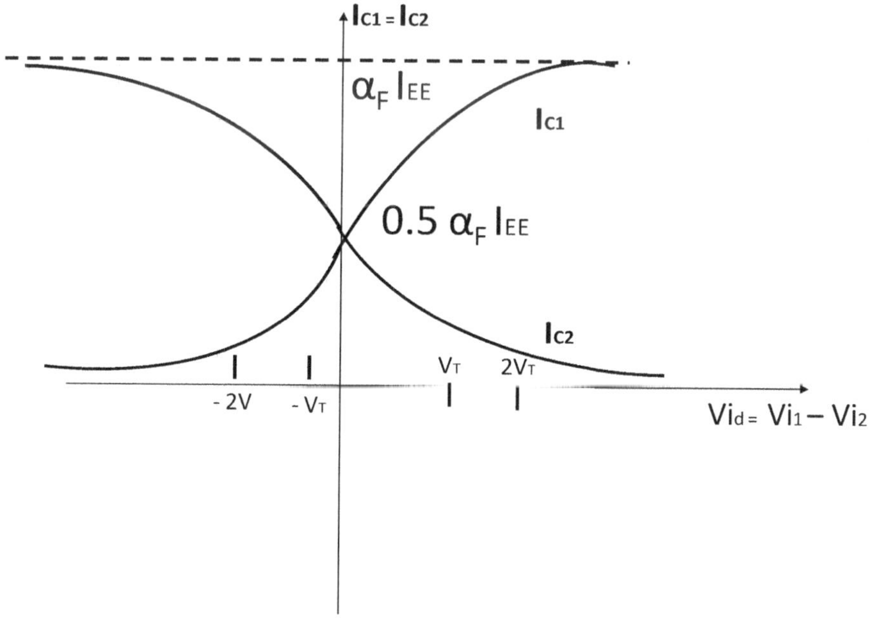

Figure 1.56. *Characteristics I_c (V_{id})*

Output voltage

The most interesting output is often the difference:

$$V_{od} = V_{o1} - V_{o2} = R_C (I_{C1} - I_{C2}),$$

$$V_{od} = \alpha_F . I_{EE} . R_C . \tanh((-Vid/V_T)$$

An advantage of the differential amplifier:

If $V_{od} = 0$, then $V_{od} = 0$.

It is therefore possible to make direct couplings of cascaded stages without introducing offsets.

Small signal analysis

The DC differential voltage is assumed to be zero.

Gain and input resistance

– The superposition theorem: it is located in the linearity zone.

– Gain in common and differential modes occurs using the half-circuit concept.

The circuit is studied in differential and common modes.

Differential mode alone

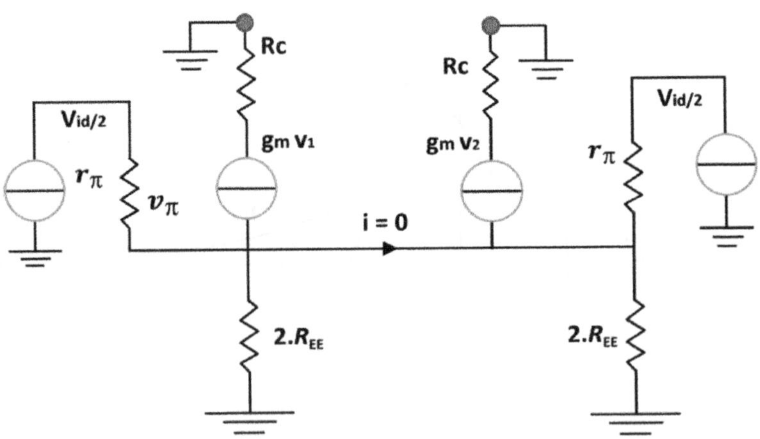

Figure 1.57. *Small signal schematic for the differential mode. For a color version of this figure, see www.iste.co.uk/gontrand/analog2.zip*

1) R_{EE} and I_{EE}: constant; so V_E constant.

Voltages of the same modulus, but of different signs, are applied to the two bases.

Since the circuit is balanced, the voltage at the emitters varies by the behavior of the circuit is therefore not affected, even if the R_{EE} circuit is shorted.

So, we can only study a half-circuit.

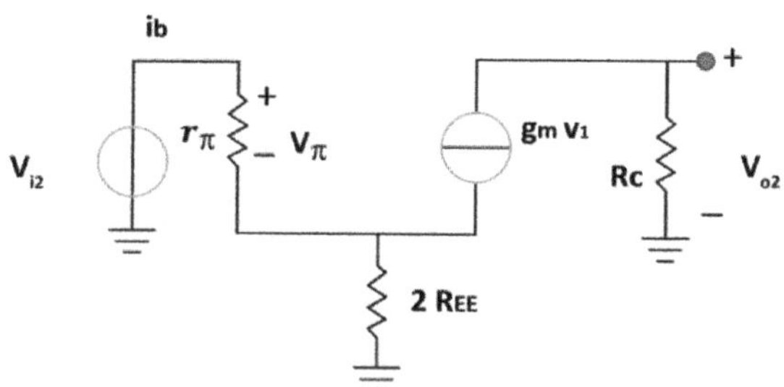

Figure 1.58. *Equivalent half circuit. For a color version of this figure, see www.iste.co.uk/gontrand/analog2.zip*

$v_0/2 = -g_m R_c v_i/2$

$$A_{dm} = \frac{v_{od}}{v_{id}} = -g_m R_c$$

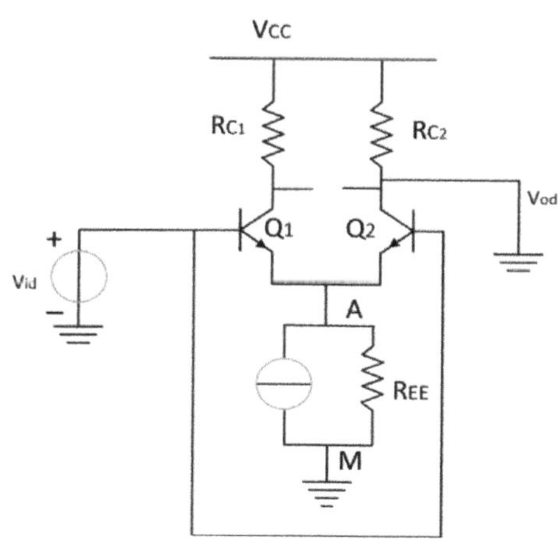

Figure 1.59. *Simplified diagram. For a color version of this figure, see www.iste.co.uk/gontrand/analog2.zip*

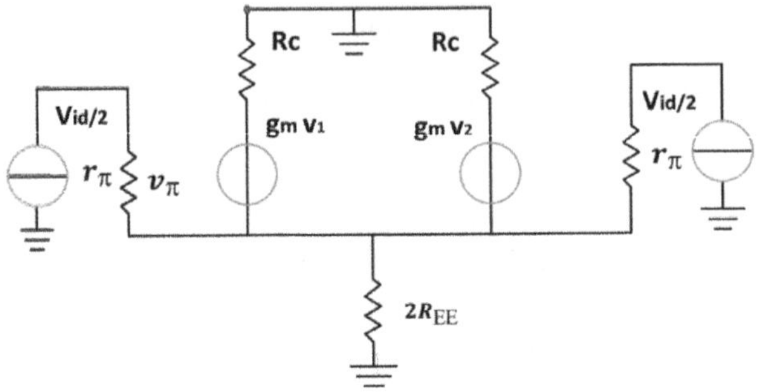

Figure 1.60. *Small signal diagram. For a color version of this figure, see www.iste.co.uk/gontrand/analog2.zip*

The two transistors have the same voltage applied across the emitter/base junction.

The symmetry of the circuit: no current in the wire joining the bases; it is therefore possible to remove the latter and study only a half-circuit.

Looped system: v = 0:

$$v_{ic} - .i_b.r_\pi - i_b.(\beta_o+1).\ 2\ R_{EE} = 0 \qquad [1.109]$$

$$v_{02} = Rc\ .i_{02} = -Rc\ .\ \beta_o\ i_b = v_{ic}.(\beta_o\ Rc/(r_\pi+2R_{EE}(\beta_o+1))$$

$$A_{cm} = \frac{V_{02}}{V_{ic}} = -\ g_m R_c/(1+2g_m R_{EE}(1/\beta_0+1)) \qquad [1.110]$$

Common mode rejection rate

Most circuits require differential voltage amplification, even in the presence of fluctuating common mode voltage.

The common mode signal produces a drowned error in the signal.

CMRR (common mode rejection rate):

$$= \frac{A_d}{A_c} = 1 + 2g_m R_{EE}(1/\beta + 1) \qquad [1.111]$$

Therefore, it is necessary to increase the output resistance of the bias current source R_{EE}.

NOTE.– This CMRR is part of the total CMRR.

Input resistance in differential and common modes

A differential amplifier often constitutes a first stage.

$$R_{id} = \frac{v_{id}}{i_b} = 2\,r_\pi \qquad [1.112]$$

r_π increases if β_0 increases and the collector current decreases.

So, if we work at low levels of bias current:

$$R_{id} = \frac{v_{ic}}{i_\beta} = r_\pi + 2R_{EE}(1+\beta_1) \qquad [1.113]$$

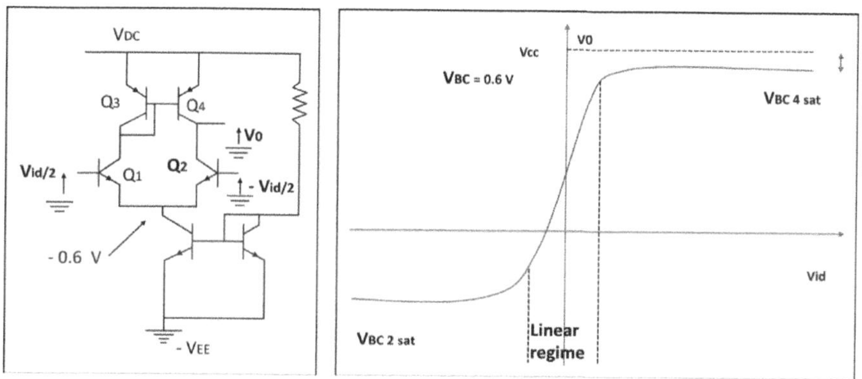

Figure 1.61. *Differential amplifier and its voltage gain. For a color version of this figure, see www.iste.co.uk/gontrand/analog2.zip*

Transfer function $V_0 = V_0\,(V_{id})$

$$I_{C1} = I_S.(1+\frac{V_{CE1}}{V_{AN}})\exp\frac{V_{BE1}}{V_T} \quad I_{C2} = I_S.(1+\frac{V_{CE2}}{V_{AN}})\exp\frac{V_{BE2}}{V_T} \quad I_{C3} = \ldots \quad I_{C4} = \ldots \qquad [1.114]$$

$$V_{id} = V_T \log\frac{I_{C1}\,(1+\frac{V_{CE1}}{V_{AN}})}{I_{C2}\,(1+\frac{V_{CC}}{V_{AN}})}\frac{I_{C2}}{I_{C1}} = \frac{I_{C4}}{I_{C3}} = \frac{(1+\frac{V_{CE4}}{V_{AN}})}{(1+\frac{V_{BE}}{V_{AP}})}$$

$$V_0 = V_{CC} - V_{BE} + \frac{\exp(\frac{V_{id}}{V_T})-1}{\frac{1}{V_{AP}}\exp(\frac{V_{id}}{V_T})+\frac{1}{V_{AN}}} = V_{CC} - V_{BE} + \frac{2.(\frac{V_{AP}.V_{AN}}{V_{AP}+V_{AN}}).\tanh(\frac{V_{id}}{2V_T})}{1+\frac{V_{AN}-V_{AP}}{V_{AN}-V_{AP}}\tanh((\frac{V_{id}}{2V_T})} \qquad [1.115]$$

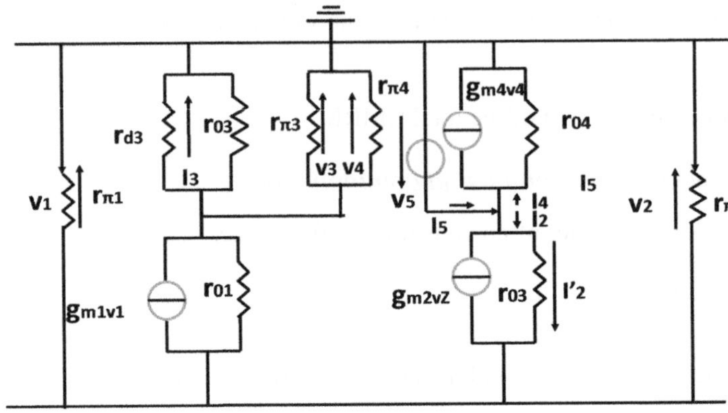

Figure 1.62. *Output resistance: equivalent diagram. For a color version of this figure, see www.iste.co.uk/gontrand/analog2.zip*

$$g_{m1}\left(\frac{v_{id}}{2} - v_E\right) + \frac{-v_3 - v_E}{r_{01}} ==> -v_E\left(\frac{1}{r_{01}} + g_{m1}\right) - v_3\left(\frac{1}{r_{01}} + \frac{1}{r_{d3}}\right) = -g_{m1}\left(\frac{v_{id}}{2}\right) \quad [1.116]$$

$$g_{m2}\left(\frac{v_{id}}{2} - v_E\right) + \frac{v_0 - v_E}{r_{02}} ==> -v_E\left(\frac{1}{r_{02}} + g_{m2}\right) - g_{m1}\, v_3\left(\frac{1}{r_{01}} + \frac{1}{r_{d3}}\right) + \left(\frac{1}{r_{01}} + \frac{1}{r_{d4}}\right) = -g_{m2}\left(\frac{v_{id}}{2}\right) \quad [1.117]$$

$$g_{m1} = g_{m2} = g_{m3} = g_{m4} = \frac{1}{r_d} \gg \frac{1}{r_{01}} + \frac{1}{r_{02}} + \frac{1}{r_{01}} + \frac{1}{r_{04}} +$$

$$R_0 = \left(\frac{v_6}{i_5}\right)$$

$$i_5 = i_2 + i_3$$

$$i_3 = -g_{m4}v_4\, i_5$$

$$i_5 = \frac{v_5}{r_{04}} = 2i_2$$

$$v_E = (\mathrm{r}_\pi // \mathrm{r}_{eq})\, i_2;\ \mathrm{r}_{eq} = \mathrm{r}_{d1}$$

$$v_5 - v_E = r_{02}(i_2 - g_{m2}v_2) = r_{02}(i_2 - g_{m2}v_E)$$

$$v_2 = (\mathrm{r}_{\pi2} // \mathrm{r}_{eq}) + r_{02}(1 + g_{m2}\, \mathrm{r}_{\pi2} // \mathrm{r}_{eq})\, i_2$$

$$r_{\pi 2} // r_{eq} = \frac{\beta r_{d2} r_{d1}}{\beta r_{d2} + r_{d1}} \neq \beta r_{d2} r_{d1} \beta r_{d2} r_{d1} = \frac{1}{g_{m2}} \qquad [1.118]$$

$$v_5 = (\, r_{d1} + 2r_{02})\, i_2 \sim 2r_{02})\, i_2$$

$$i_5 = \frac{v_5}{r_{04}} + \frac{v_5}{r_{02}}$$

$$R_0 = r_{02} // r_{04}$$

Here, collector resistors are active charges.

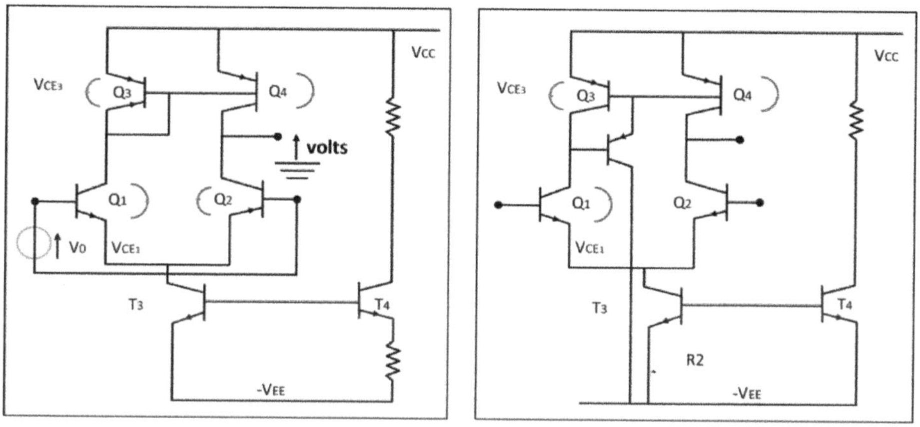

Figure 1.63. *Active charge differential amplifiers. For a color version of this figure, see www.iste.co.uk/gontrand/analog2.zip*

$$\neq \beta r_{d2} r_{d1} \beta r_{d2} r_{d1}$$

$$3 = V_{CE3}\, V_{CE1}$$
$$V_{CE1} = V_{CE2}$$

$$I_{CE2} = -\, I_{CE4} = -\, I_{CE3}(\frac{I_{S2}}{I_{S3}})$$

$$I_{C1} = -\, I_{C3}(1 + \frac{2}{\beta_p})$$

$$V_0 = V_{BE1} - V_{BE2} = V_T Log(\frac{I_{C2}}{I_{C3}} . \frac{I_{S2}}{I_{S1}})$$

$$V_0 = V_T Log(\frac{I_{S3}}{I_{S4}} . \frac{I_{S2}}{I_{S1}} (1 + \frac{2}{\beta_p}) \qquad [1.119]$$

$$I_{Sp} = \frac{I_{S3} + I_{S4}}{2}$$
$$I_{S4} = \frac{I_{S1} + I_{S2}}{2}$$

$$\Delta_{Sp} = I_{S3} - I_{S4}$$
$$\Delta_{Spddd} = I_{S1} - I_{S2}$$

$$V_0 = V_T\left(\frac{\Delta_{Sp}}{I_{Sp}} - \frac{\Delta_{SN}}{I_{SN}} + \frac{2}{\beta_P}\right) \qquad [1.120]$$

Output stage

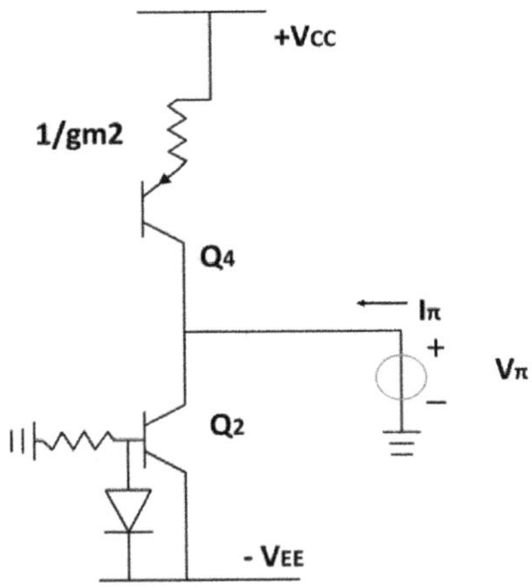

Figure 1.64. *Output stage. For a color version of this figure, see www.iste.co.uk/gontrand/analog2.zip*

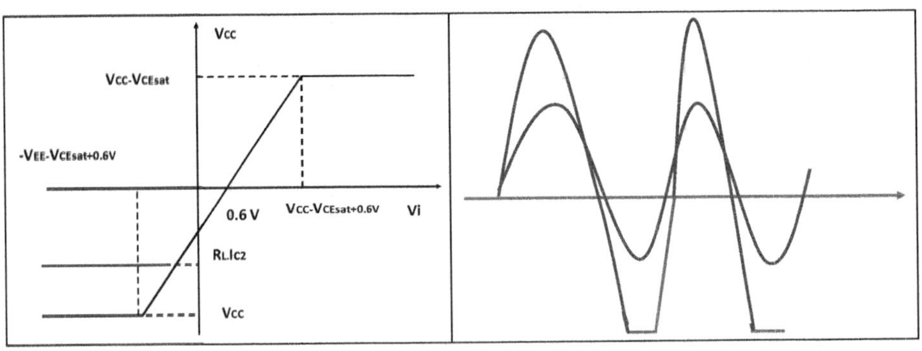

Figure 1.65. *Output saturation. For a color version of this figure, see www.iste.co.uk/gontrand/analog2.zip*

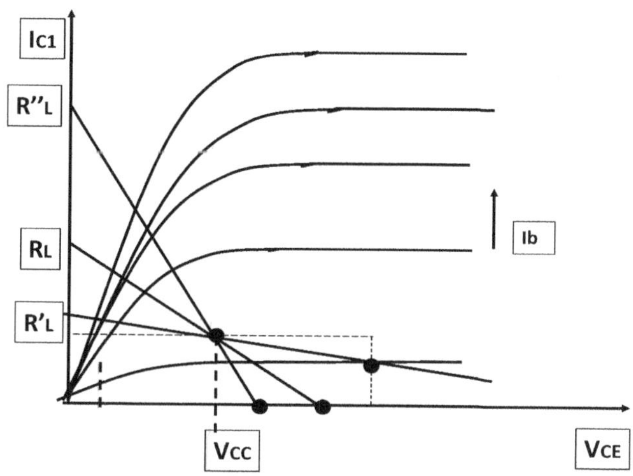

Figure 1.66. *Early curves and charge lines: $R_L > R'_L > R''_L$.*

NOTE.– Early curves.

The real characteristics I_c (V_{ce}) do not saturate, but are tilted, converging to the left on the point: $-V_A$. The corresponding ideal curves tend toward "infinity".

Indeed, when V_e increases, the width of the neutral base decreases and the V_{CB} desertion zone increases; thus, the gradient of the profile of the minority in the base, and therefore the diffusion current, is increased.

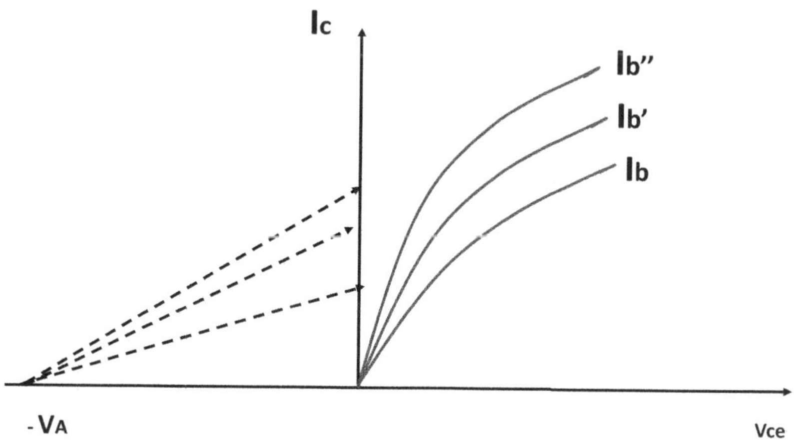

Figure 1.67. *Early curves. For a color version of this figure, see www.iste.co.uk/gontrand/analog2.zip*

$$r_0 = \frac{\Delta I_C}{\Delta V_{CE}} = \frac{V_A}{I_C},$$

$$I_C = I_S\ (1 + \frac{V_{CE}}{V_A})exp\frac{V_{BE}}{V_T} \quad [1.121]$$

Let us go back to the common emitter.

Charge lines

$$R_L = \frac{V_{cc} - V_{E2sat}}{I_{C2}}$$

$$P_m = \frac{1}{2}\frac{V_{cc} - V_{E2sat}}{I_{C2}}$$

Efficiency:

$$P_m = \frac{P_m}{(V_{CC}+V_{EE}).I_{C2}} = \frac{Power\ delivered\ to\ the\ charge}{Source\ power} \quad [1.122]$$

Dissipated power in transistor Q_1

$$I_{C1} = \frac{P}{V_{cE_1}}$$

$$\frac{dI_{C1}}{dV_{CE_1}} = -\frac{P}{V^2{}_{CE_1}} = -\frac{I_{C1}}{V_{CE_1}} = -\frac{I_{C1}}{V_{CC}} = -\frac{1}{R_L}$$

Push-pull assembly

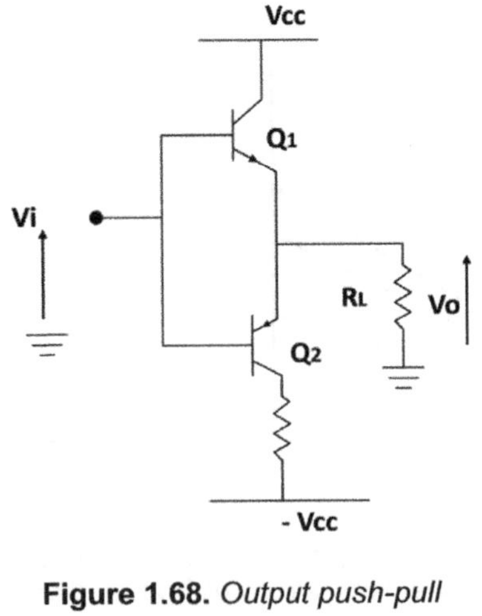

Figure 1.68. *Output push-pull*

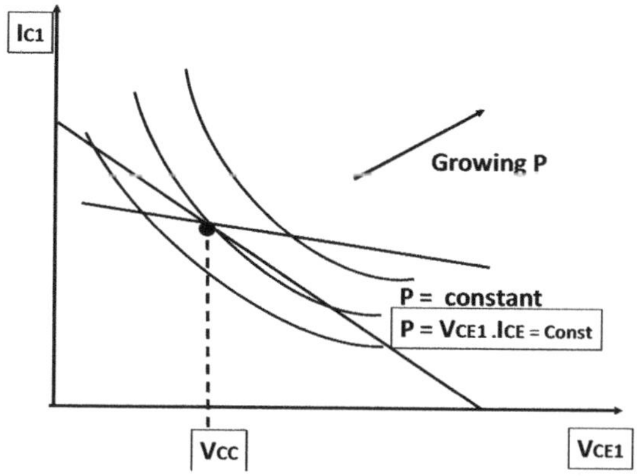

Figure 1.69. *Power constraints*

A poor choice of the charge resistor can lead to a very high power being dissipated in the transistor, thereby destroying the output stage of the amplifier.

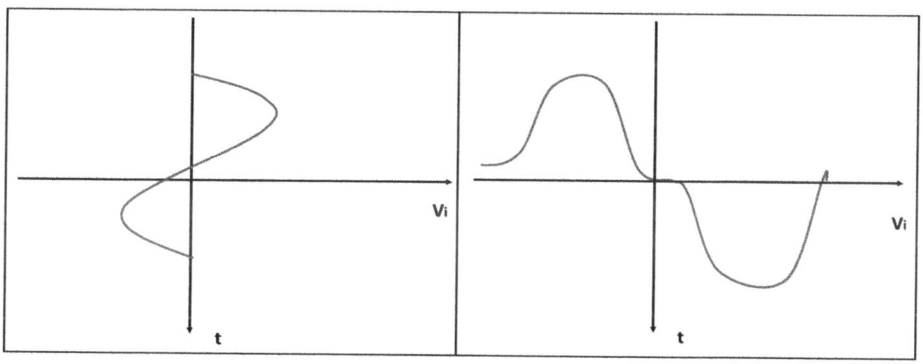

Figure 1.70. *Input and output. For a color version of this figure, see www.iste.co.uk/gontrand/analog2.zip*

1.8. Basic analog functional blocks in (C)MOS technology

We propose studying two elementary configurations most commonly used in analog CMOS circuitry: common source mounting and cascode mounting.

1.8.1. *Common source NMOS transistors*

The following assembly is carried out:

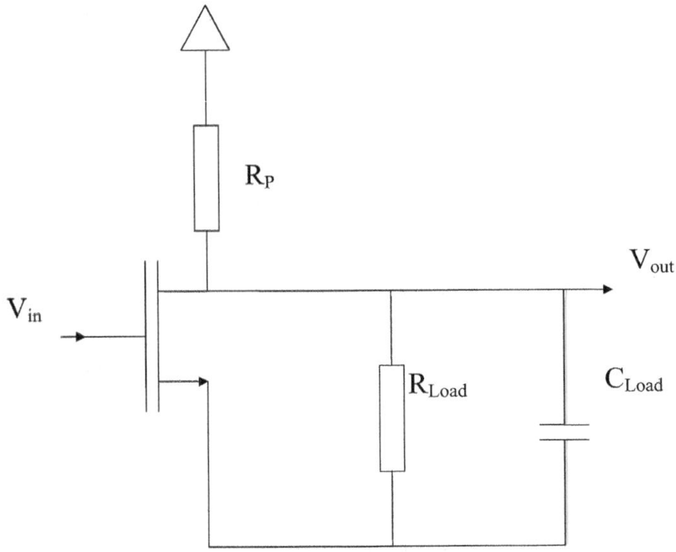

Figure 1.71. *Common source mounting*

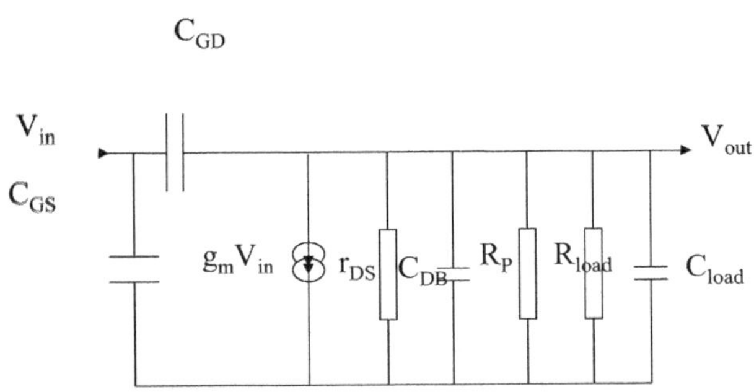

Figure 1.72. *Equivalent scheme in small signals*

We have: $i_{cc0} = -g_m V_{in}$: short-circuit current without considering capacitances; indeed, capacitors have high impedance at low frequencies.

The impedance, after cancelling the input sources, seen from the (node) output (when $V_{in} = 0$) is:

$$Z_{out} = R_{LT} = r_{DS} // R_{load} // R_P \quad [1.123]$$

Hence, the low frequency gain is:

$$Av_0 = \frac{v_{out}}{v_{in}} = \frac{Z_{out}\ i_{cc0}}{v_{in}} = -g_m R_{LT} \qquad [1.124]$$

Let R_L be the output resistance of the transistor alone, and it is $R_L = R_{load} // R_P$:

$$\Rightarrow Av_0 = -g_m R_L // r_{DS} \qquad [1.125]$$

NOTE.– We will insist somewhat on this simple, but realistic, example of the MOS in common source, to show a general methodology; of course, the goal is not to develop exhaustive calculations for other mountings developed in many fundamental books.

In fact, the approach is to locate the poles and zeros of a transfer function.

The representation of the modulus of the gain at low frequency as a function of R_L in log-log scale (Figure 1.73) shows that the maximum value reached by this modulus is the product: $g_m r_{DS}$.

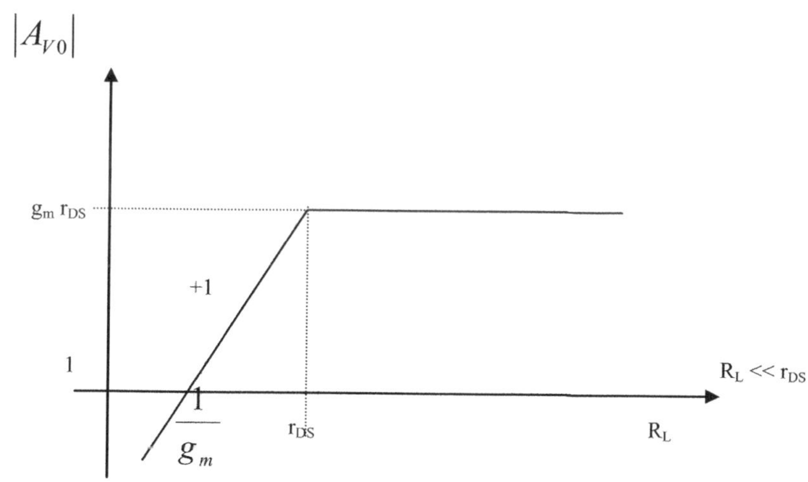

Figure 1.73. *Gain module*

Let us now take capacitors into account:

$$i_{cc} = -g_m v_{in} + \frac{v_{in}}{Z_{Cgd}}$$

that is:

$$i_{cc} = -g_m v_{in}\left(1 - \frac{C_{gd}}{g_m}p\right) \quad [1.126]$$

But $Z_{out} = r_{DS}//R_{load}//R_p//Z_{(C_{DB}+C_{GD}+C_{load})}$

So:

$$Z_{out} = \frac{R_{LT}}{1 + R_{LT}\,C_{LT}\,p} \quad [1.127]$$

where:

$C_{LT} = C_{DB} + C_{GD} + C_{Load}$

Hence, the voltage gain, as a function of frequency, is given as:

$$A_v = \frac{v_{out}}{v_{in}} = \frac{Z_{out}\,.i_{cc}}{v_{in}} = g_m R_{LT}\,.\frac{\left(1 - \frac{C_{gd}}{g_m}p\right)}{1 + R_{LT}\,C_{LT}\,p} \quad [1.128]$$

C_{GD} creates a positive zero at frequency: $g_m/2\pi X_{GD}$.

With C_{DB} and C_{Load} capacitances, a negative pole at the frequency is created: $1/(2pR_{LT}C_{LT})$.

Figure 1.74 shows an asymptotic log-log diagram of the evolution of the value of the frequency of this pole and of this zero as a function of R_L.

In Figure 1.75, R_L is plotted on the ordinate and the frequency on the *Y*-axis; the locus of the poles and zeros is obtained as a function of R_L.

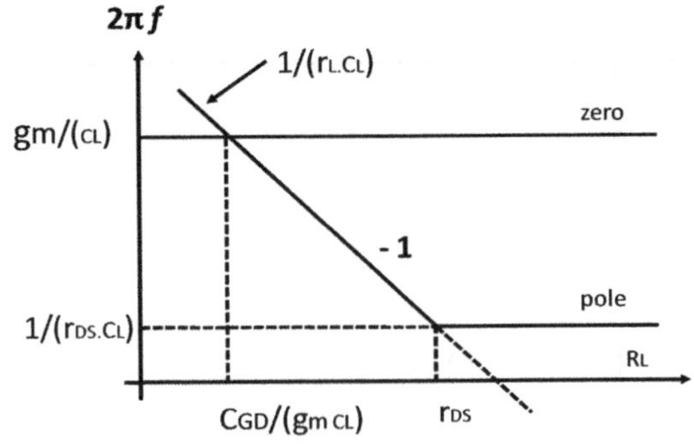

Figure 1.74. *Frequency versus R_L*

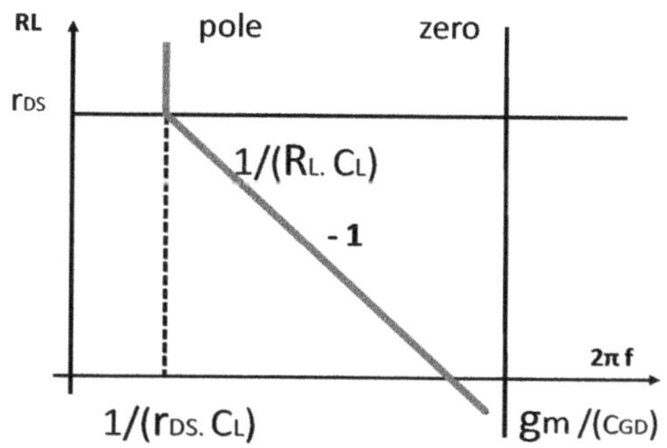

Figure 1.75. *Location of pole and zero as a function of R_L. Asymptotic log-log diagram. For a color version of this figure, see www.iste.co.uk/gontrand/analog2.zip*

Using the two graphs (Figures 1.73 and 1.75), it is easy to plot the Bode diagram of the voltage gain modulus, with R_L as a parameter.

NOTE.– Using two or more graphs to build another one is highly appreciated by specialists in the field; we avoid calculations as much as possible.

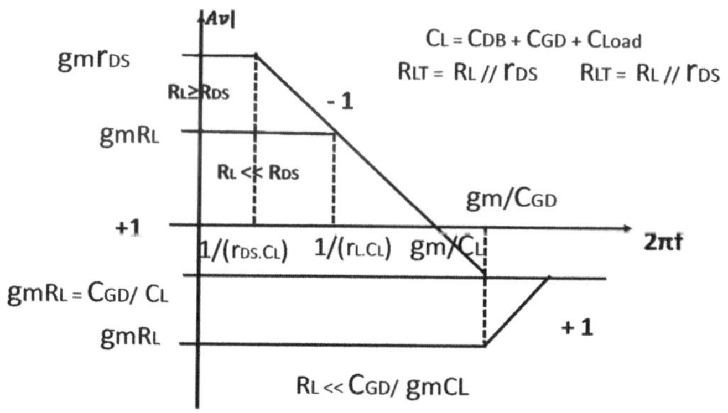

Figure 1.76. *Asymptotic Bode Diagram of the voltage gain module of Figure 1.71 as a function of frequency and for different values of R_L. For a color version of this figure, see www.iste.co.uk/gontrand/analog2.zip*

In the most frequent case where the modulus of Av >1 and where the frequency of the pole is clearly lower than that of the zero, it is noted that the frequency at unit gain is independent of R_L and is equal to g_m/C_L. If the capacitance C_{LOAD} is negligible compared to $C_{DB}+C_{GD}$, the unit gain frequency is equal to $g_m(C_{DB}+C_{GD})^{-1}$

and depends only on the transistor alone; this is typically a merit factor of the transistor.

The effect of the capacitance C_{GD} is not limited to the appearance of zero and the participation, in the pole, of the voltage gain expression of the common source circuit. Indeed, calculation of the input impedance of the equivalent diagram reveals a capacitive component C_{GDin}, due to C_{GD}, expressed as follows:

$$C_{GDin} = (1 + g_m R_{LT}).C_{GD} = (1 - A_{V0}).C_{GD} = |A_{VD}|C_{GD} \quad [1.129]$$

This is the famous Miller effect: any impedance connecting the input and output of an amplifier is multiplied by the gain if brought back to the input.

NOTE.– The latter is widely used in integrated circuits to reduce the place of capacitors.

The influence of the C_{GDin} capacitance can become dominant. As shown in Figure 1.77, if voltage is driven through an output resistor Rs (the output impedance of the input generator) and the gain Gv is defined as the ratio V_{out}/V_B (and not Vout/Vin, previously done), it is shown, without approximation, that the gain is expressed as:

$$G_V = -g_m R_{LT} \cdot \frac{(1-\frac{C_{GD}}{g_m}p)}{1+(R_S C_{GS} + (R_{LT} + R_S)(1+g_m R_{LT}))C_{GD} + R_{LT} C_L)p] + (R_L R_S C_{\Sigma\pi})p^2} \quad [1.130]$$

where $C_{\Sigma\pi} = C_L C_{GB} + C_L C_{GD} + C_{GD} C_{GS}$

In addition, the input capacity is expressed as:

$$C_{GDin} = (R_{LT}/R_S + 1 + g_m R_L)C_{GD} \quad [1.131]$$

When the resistance Rs is zero, gain Av calculated in Figure 1.90 is found again.

If capacitance C_{GD} is zero, it is found that capacitance C_L creates a pole at the angular frequency $1/R_{LT}.C_L$ (pole already known from the output impedance) and that capacitance CGS associated with the resistor R_S creates a pole at the angular frequency $1/R_{LT}.C_L$ (pole of the short-circuit current):

$$v_{IN} = v_B/(1+R_S C_{GB}) \quad [1.132]$$

If the capacitance becomes large enough to ignore other capacitances, we get:

$$C_{\Sigma\pi} = C_{GD}(C_L + C_{GS}) \quad [1.133]$$

and:

$$(R_S\, C_{GS} + (R_{LT} + R_S)(1 + g_m R_{LT}))\, C_{GD} + R_{LT}) = (R_{LT} + R_S(1 + g_m R_{LT}))\, C_{GD} = R_S\, C_{GDin} \qquad [1.134]$$

The pole due to the action of C_{GDin} becomes dominant; the term ($1+R_S\,C_{GDin}.p$) can thus be put into factor in the denominator of the G_V expression, the latter becoming:

$$G_{vV} = -g_m R_{LT} \cdot \frac{(1-\frac{C_{GD}}{g_m}p)}{(1+R_S\,C_{GDin}\,p)(1+\frac{R_{LT}\,R_S\,(C_L+C_{GS})}{R_{LT}+R_S\,(1+g_m R_{LT})}p} \qquad [1.135]$$

If the gain LF is of high value, we find $C_{GDin} = C_{GD} = g_m\, R_{LT}.C_{GD}$ and we have:

$$\frac{R_{LT}\,R_S\,(C_L+C_{GS})}{R_{LT}+R_S(1+g_m R_{LT})} = \frac{C_L+C_{GS}}{g_m} \qquad [1.136]$$

The voltage gain G_V can then be written as:

$$G_{vV} = -g_m R_{LT} \cdot \frac{(1-\frac{C_{gd}}{g_m}p)}{(1+R_S\,g_m\,R_{LT}\,C_{GD}\,p)(1+\frac{(C_L+C_{GS})}{g_m}p} \qquad [1.137]$$

It can be seen that the second nondominant pole is a pulse $g_m/(C_L+C_{GS})$, which can be greater than $1/(R_{LT}C_L)$ and $1/(R_SC_{GS})$ pulses of poles when $C_{GD} = 0$.

Consequently, when capacitance C_{GD} increases, the dominant pole and the zero are brought back with a –1 slope toward low frequencies, while the dominant pole is shifted toward high frequencies (Figure 1.77); this is pole splitting which is often used for frequency compensation.

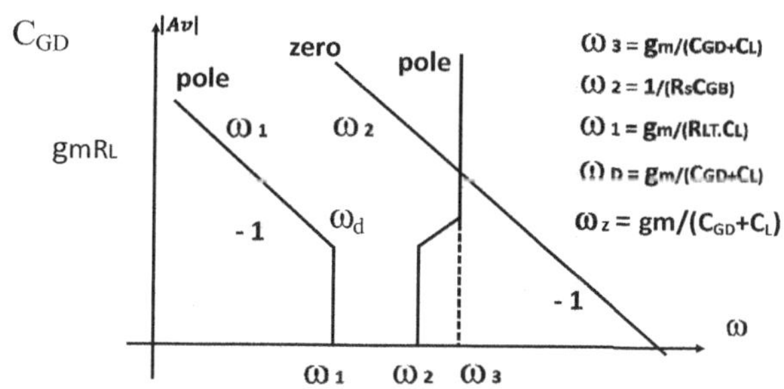

Figure 1.77. *Effect of the separation of poles on the location of poles and zeros of the voltage gain expression Gv of the common source mounting as a function of the capacitance C_{GD} (on the ordinate)*

1.8.1.1. *Cascode assembly*

The low-frequency voltage gain of a common source circuit is limited by the term g_m r_{DS} (for infinite R_L). For equal transconductance, it is the output resistor r_{DS} that sets the maximum value of the output impedance, and therefore of the gain, at low frequencies. The maximum gain of a common source amplifier in CMOS technology is a few tens. In order to obtain a higher amplification, the output impedance of the circuit is sought to be increased due to the driving transistor M_1; for this purpose, a second cascode-connected transistor is added (Figure 1.78).

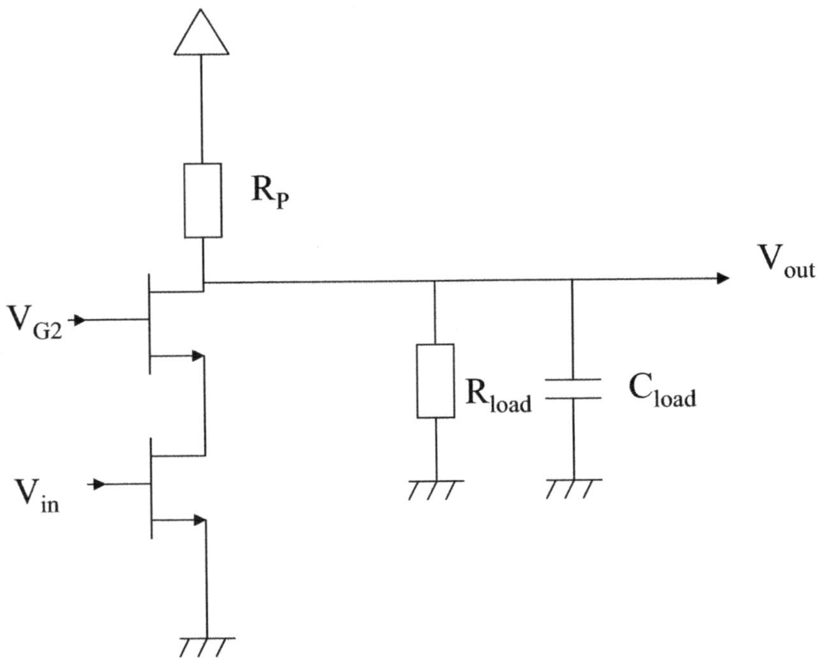

Figure 1.78. *Cascode assembly*

$$V_{GS2} = V_G - V_{S2} = -V_S \ (V_G = 0)$$

The equivalent schematic for small alternating signals at low frequencies is as follows:

The output short-circuit current is obtained from the preceding diagram by connecting the output node to ground.

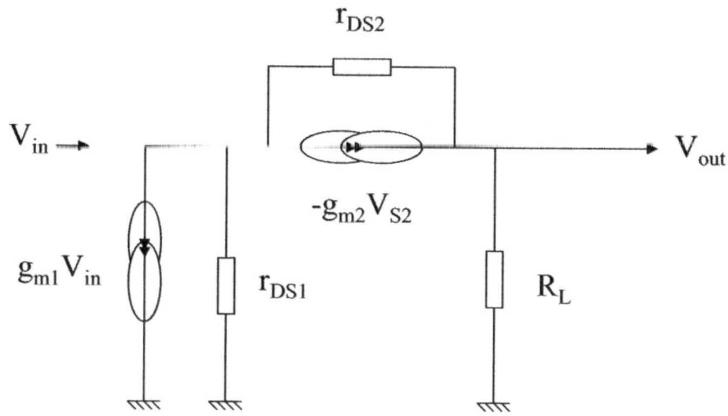

Figure 1.79. *Small signal equivalent schematic (R_L = Rp // R_L)*

And we find:

$$i_{cc0} = g_{m2}\, v_{S2} + v_{S2/}\, R_{Ds2} \text{ and } i_{cc0} = -g_{m1}\, V_{in} - v_{S2/}\, R_{Ds2}$$

so:

$$i_{cc0} = -g_{m2}\, V_{in} \frac{(1 + g_{m2}\, r_{DS2})r_{DS1}}{r_{DS2} + (1 + g_{m2} r_{DS2})r_{DS1}}$$

or:

$$i_{cc0} = -g_{m2}\, V_{in} \frac{\dfrac{1}{r_{DS2}} + g_{m2}}{\dfrac{1}{r_{DS1}} + \dfrac{1}{r_{DS2}} + g_{m2}} = -g_{m2}\, V_{in} \frac{r_{DS1} // r_{DS2} // \dfrac{1}{g_{m2}}}{r_{DS2} // \dfrac{1}{g_{m2}}} \qquad [1.138]$$

which can be interpreted as follows: short-circuited and seen from the source of transistor (cascode) M_2, and current source $g_{m2}V_{S2}$ behaves like a value resistor.

$$\frac{1}{g_{m2}}$$

The drive current $g_{m1}V_{in}$ is then shared between resistors r_{DS1}, r_{DS2} and $\dfrac{1}{g_{m2}}$.

The short-circuit current at the output is the part of this driving current that flows through the resistors $\frac{1}{g_{m2}}$ and r_{DS2}. We actually have a simple current divider.

In the case where all the transistors are in saturation mode, we have $r_{DS} >> \frac{1}{g_m}$, which results in $i_{cc0} \approx - g_{m1} V_{in}$, value obtained in common source assembly.

With the help of Figure 1.80, the output impedance is obtained in small alternating signals.

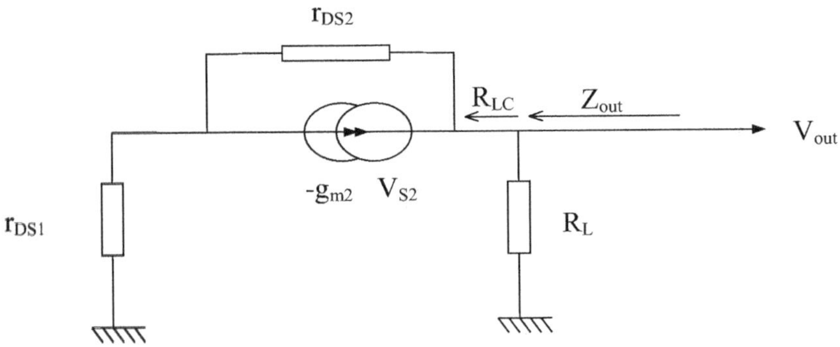

Figure 1.80. *Small signal diagram to calculate the output impedance of Figure 1.79*

$$Z_{out} = R_L // R_{LC}$$

A simple calculation shows that:

$$R_{LC} = r_{DS2} + (1 + g_{m2}\, r_{DS2})\, r_{DS1} \quad [1.139]$$

This is where the first cascode effect comes in: the impedance r_{DS1} seen from the output is multiplied by the factor $(1 + g_{m2}\, r_{DS2})$ in the saturation zone, this term g_{m2} r_{DS2} can be worth a few tens. Consequently, the R_{LC} output impedance seen toward the drive transistor is one or two orders of magnitude greater than the r_{DS1} output impedance of a common source circuit.

The low-frequency voltage gain for small alternating signals (Figure 1.79) is therefore:

$$A_{v0} = \frac{z_{out} i_{cc0}}{v_{in}} = -g_{m2}\frac{r_{DS1} // r_{DS2} // \frac{1}{g_{m2}}}{r_{DS2} // \frac{1}{g_{m2}}}(R_L // R_{LC}) \qquad [1.140]$$

or even:

$$A_{Vo} = -g_{m1}\frac{(1+g_{m1}r_{DS2})r_{DS1}R_L}{R_L + r_{DS2} + (1+g_{m2}r_{DS2})r_{DS1}} \qquad [1.141]$$

If $g_{m2}\, r_{DS2} >> 1$, we find $A_{v0} \approx -g_{m1}\,(R_L // R_{LC})$

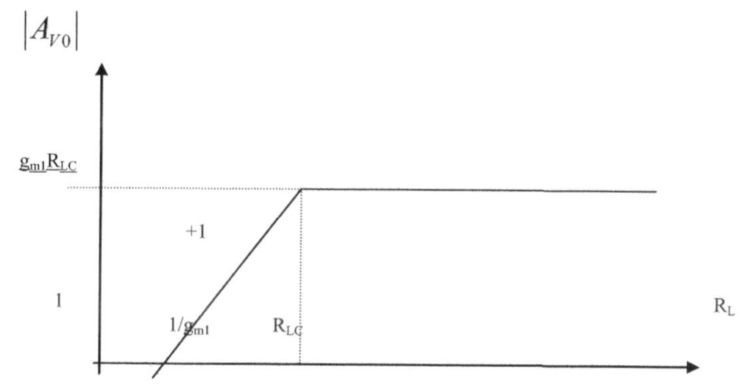

Figure 1.81. *Diagram log–asymptotic log of the module of the voltage gain BF of the diagram of Figure 1.78 as a function of R_L*

Conclusion on the two assemblies

In order to qualitatively compare the voltage gain LF of the cascode circuit to that of the common source, with identical drive transistors (g_{m1} and r_{DS1}) and as a function of the charge resistance R_L, we calculate the ratio:

$$\frac{A_{V0cas}}{A_{V0SC}} \approx \frac{(R_L // R_{LC})}{(R_L // r_{DS1})} = \frac{1+\frac{R_L}{r_{DS1}}}{1+\frac{R_L}{R_{LC}}} \qquad [1.142]$$

The asymptotic log-log diagram (Figure 1.78) of this expression (with $r_{DS1} < R_{LC}$) is:

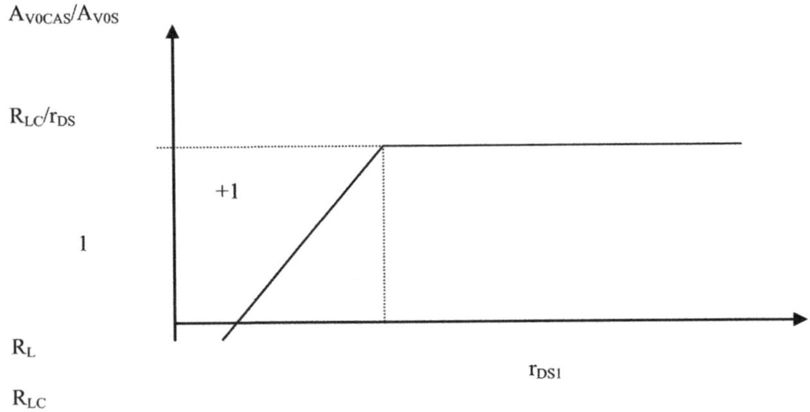

Figure 1.82. *Diagram log–asymptotic log of the module of the voltage gains LF of the diagram of Figure 1.78 as a function of R_L*

For a charge R_L low in ohms ($R_L < r_{DS1}$), the introduction of a transistor (cascode) does not improve the low-frequency voltage gain of a common source circuit. On the other hand, if the charge R_L is greater than the impedance (cascode) R_{LC}, the gain is improved by the ratio $\frac{R_{LC}}{r_{DS1}}$ of about $g_{m2}\, r_{DS2}$ (which corresponds to the increase in the output impedance seen toward the driving transistor M_1).

As for the commuted source, here we can study the frequency (wobulation). Calculations are rather heavy; but the methodology is the same.

1.8.2. *Reminder on the general structure of the operational amplifier*

The operational amplifier is the basic circuit of analog. We will talk (again) essentially about the input stage and its study in common mode (the differential mode being systematically treated in the academic literature). The aim here is somewhat to understand the operation of a circuit and the operating regimes of the devices constituting it.

Characteristics and performance of the operational amplifier (O.A.)

The operational amplifier is defined as a voltage source controlled by a differential input voltage. The ratio between the variation of the output voltage and the variation of the differential input voltage defines the differential gain: A_{VD}, main characteristic of the O.A.

In order to achieve the desired function, O.A.s are in most cases used in a closed-loop system making it possible to control the output voltage as a function of the input voltage.

The voltage gain, in open loop, in other words the transfer function of the O.A., must then be as large as possible, so that the gain in closed loop depends only on the feedback or as little as possible on the gain in open loop.

Under conditions of stability and for certain physical reasons for which the system has a transfer function varying (with a slope − 1) from the frequency of the dominant pole up to the unit gain frequency: the gain-band, open-loop and closed-loop products are equal.

The output voltage of the O.A. must depend as little as possible on the input voltage. The common mode gain A_{VMC}, the ratio of the variation of the output voltage to the variation of the common mode input voltage, and the common mode rejection ratio (CMRR), the ratio of the differential gain to the common mode gain, are defined.

The O.A.s are generally used in a linear and counter-reaction regime. In order to limit the functioning in these regions, we mention some important notions such as maximum rate of variation of the output voltage (slew rate), the harmonic distortion brought by the O.A. to the output signal, the dynamic operating range for the common mode input voltage CMR (common mode range) and the dynamic output range, an extreme value that the output voltage can take.

If the CMR of an O.A. varies from negative V_{SS} to positive V_{DD}, this O.A. is said to have a rail-to-rail input mode; likewise, for the output if the output dynamic range extends from Vss to V_{DD}.

Summary

A well-performing O.A. is mainly characterized by:

– high A_{VD0} low-frequency differential gain;

– a frequency at the highest possible unit gain B_1 while retaining a slope of order –1 for the transfer function between the frequency of the dominant pole and the frequency at the unit gain;

– a high common mode rejection rate over the entire frequency range of use;

– a significant slew rate;

– low harmonic distortion at the output for a sinusoidal input signal;

– a dynamic range of use in input and output as wide as possible (ideally from V_{SS} to V_{DD}, that is, rail to rail mode);

– in addition, to approximate the characteristics of an ideal controlled voltage source, an O.A. must have the highest input impedance and the lowest output impedance. All of these characteristics must be ensured with low power consumption.

Structure of O.A.s

The basic block diagram of the O.A. structure, on which three essential functions appear, is as follows:

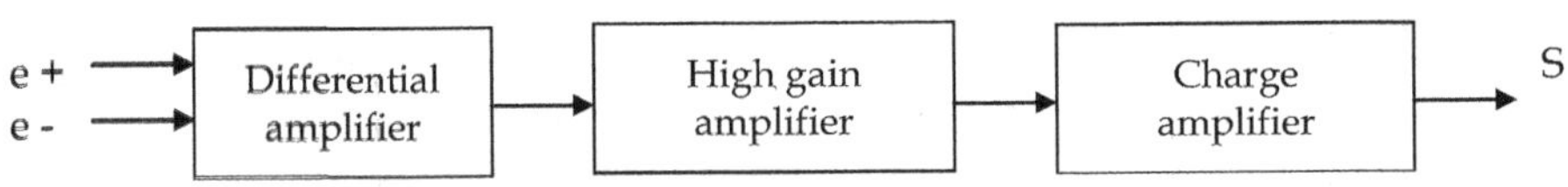

Figure 1.83. *Block diagram of the principle of the O.A*

The term "two-stage operational amplifier" is used if the input stage and possibly the gain stage contribute to the overall voltage gain of the O.A.

The input stage consists of a differential amplifier supplying to the circuit the inverted and non-inverted inputs, the large CMRR, and the high input impedance.

The second stage is a very high gain voltage amplifier, the most necessary for the whole of the open-loop O.A.

To develop a large gain, the output impedance of the preceding stage is very high. Consequently, it is necessary to have a third and final stage, the output stage, giving a low output impedance of the O.A. while ensuring a high input impedance for the gain stage.

Most of the O.A. intended for high-frequency amplifications comprise, in fact, three stages having voltage gain. The overall gain, being provided by these three stages, therefore makes it possible to reduce the gain of each, which has the advantage of improving the slew rate (by reducing the transconductance of the input stage for example) or of increasing the frequency to the unit gain, since each stage has a high passband.

Obviously, there are innumerable variants on the internal structure of the O.A. (number of stages, complexity of compensation networks, completely differential operational amplifiers, common mode feedback circuit, etc.) but the basic principles always go back to those we have just presented.

Rail-to-rail input stage

1.8.2.1. *Definition*

We want to design a differential input stage that works for common mode transistors ranging from the negative V_{SS} power supply to the positive V_{DD} (rail to rail). The dynamic common mode range CMR is then centered on the half-sum of the supply voltages which is, by definition, the analog ground. The main performances are a high differential gain and bandwidth, which will be as constant as possible over the entire CMR and a low common mode gain.

1.8.2.2. *Study of the dynamic range of common mode input voltage: MOS*

The operation of the differential stages of the NMOS type and of the PMOS type (Figure 1.84) depends on the bias of the transistors constituting them. This polarization itself depends on the input voltages and in particular on the common mode voltage.

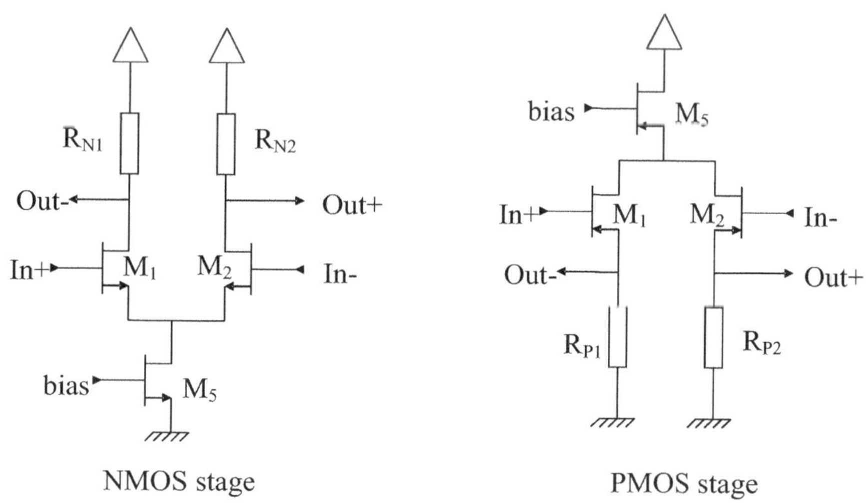

Figure 1.84. *Entry stage of an O.A*

Here:

M_5: current source;

M_1, M_2: differential pair;

R_1 and R_2: active charges.

Let us study the evolution of the amplitude of the current variation i, for the same differential excitation V_{diff}, at the input (that is to say of constant amplitude whatever the common mode voltage), as a function of the common mode voltage V_{MC}. The latter will be between V_{SS} and V_{DD}.

Based on the equivalent scheme of the differential pair (Figure 1.84), the low frequency differential gain of the input stage is calculated.

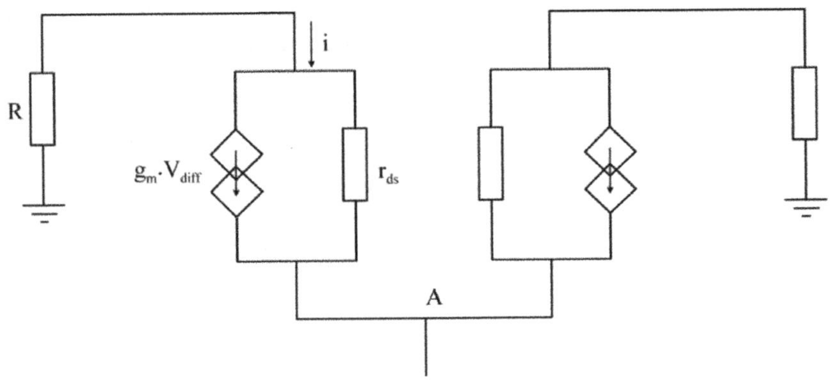

Figure 1.85. *Equivalent diagram of the differential pair*

R is the equivalent resistance of the diode-connected transistors.

We find:

$$i = g_m \cdot \left(\frac{r_{DS}}{r_{DS} + R_N} \right) \cdot v_{diff} \quad [1.143]$$

1.8.2.3. *Mode of operation of an NMOS differential stage*

The amplifier is used in common mode. Four operating zones can be distinguished (Figure 1.86), for the diagram in Figure 1.85.

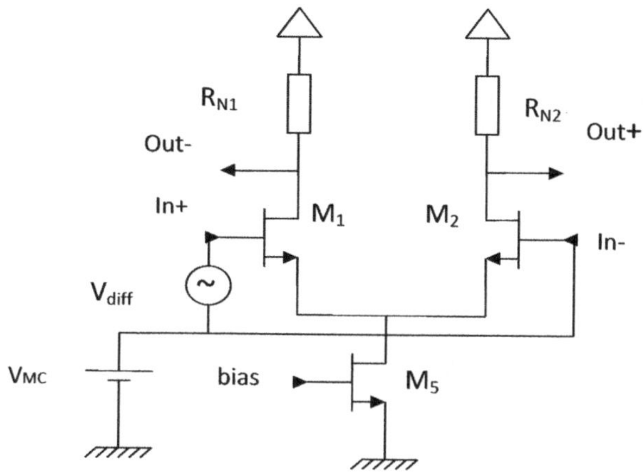

Figure 1.86. *NMOS differential stage: common mode*

Common mode voltage from V_{SS} to V_{SS} + $V_{th}(M_1)$

When the common mode voltage is less than $V_{th}(M_1)$, transistors M_1 and M_2 are turned off and no current flows in the differential stage. The drain potentials are then equal to V_{DD}, and the source potential is equal to V_{SS}.

Common mode voltage between V_{SS} + $V_{th}(M_1)$ and V_{SS} +$V_{th}(M_1)$ + $V_{dsat}(M_5)$

When the common mode voltage is between V_{SS} + Vth(M_1) and V_{SS} + $V_{th}(M_1)$+$V_{dsat}(M_1)$, the voltage V_{DS} of the transistor M_5 is less than V_{dsat} and M_5 is in the ohmic zone. Then the DC currents increase with the common mode voltage (see characteristic $I_D(V_{DS})$ of the MOS). Voltages V_{GS} of transistors M_1 and M_2 increase with the drain currents while their drain potentials decrease (the current increases in the transistorized charges R_1 and R_2).

In the ohmic zone for M_5, the transconductance of the input transistors is therefore not constant since the drain current varies. As a result, the current variation ($g_m v_{diff}$), due to the same variation of the differential input voltage, increases (in absolute value) when the common mode voltage varies from V_{SS} + $V_{th}(M_1)$ and V_{SS} + $V_{th}(M_1)$ + $V_{dsat}(M_5)$.

Common mode voltage between V_{SS} + $V_{th}(M_1)$ + $V_{dsat}(M_5)$ and V_{DD} – [$V_{dsat}(M_1)$ + $R.I_0$ – $V_{GS0}(M_1)$]

In order for transistors of the differential stage all to have reached their saturation zone, the common mode voltage at the input must have exceeded the value V_{SS} + $V_{th}(M_1)$ + $V_{dsat}(M_5)$. Under these conditions, the currents in transistors M_1 and

M_2 are constant and equal to I_0. Then voltages V_{GS} of M_1 and M_2 are constant (V_{GS0}), and they adapt to allow current I_0 to flow in the channel. Consequently, the source potential follows the variations of the gate common mode potential. The drain potentials of these input transistors are constant and are equal to V_{DD}-$R.I_0$, R being the transistorized resistor for current I_0.

In the saturation zone, at constant current and constant voltage V_{GS}, the transconductance g_m of input transistors is constant. Consequently, the current variation i due to the same variation of the input differential voltage is constant when the common mode voltage varies between $V_{SS} + V_{th}(M_1) + V_{dsat}(M_5)$ and $V_{DD} - [V_{dsat}(M_1) + R.I_0 - V_{GS0}(M_1)]$.

Common mode voltage between $V_{DD} - [V_{dsat}(M_1) + R.I_0 - V_{GS0}(M_1)]$ and V_{DD}

Finally, when the common mode voltage approaches V_{DD}, currents of transistors remain constant, imposed by M_5, but voltages V_{DS} of transistors become lower than V_{DSat}, since drain voltages remain constant and the common source potential increases. Then transistors M_1 and M_2, from the common mode voltage equal to $V_{DD} - [V_{dsat}(M_1) + R.I_0 - V_{GS0}(M_1)]$, enter a constant current ohmic zone, where, in order to compensate for a decrease in V_{DS}, V_{GS} must increase. As a result, the common source potential of M_1 and M_2 increases less rapidly than that of the common gate mode, and it stabilizes.

In a constant current ohmic zone, the transconductance g_m of the input transistors decreases when the common mode voltage varies between $V_{DD} - [V_{dsat}(M_1) + R.I_0 - V_{GS0}(M_1)]$ and V_{DD}. The amplitude of the current variation i decreases faster than the transconductance g_m since the resistance r_{DS} of the transistor M_1 also decreases and:

$$i = \left(\frac{rDS}{rDS+RN}\right) g_m v_{diff} \quad [1.144]$$

It should be noted that this last operating zone exists only if $V_{dsat}(M_1) - R.I_0 > V_{GS0}(M_1)$.

Total rail-to-rail differential structure

In CMOS technology, it is customary to replace the resistive charges R_1 and R_2, with PMOS transistors mounted as diodes M_3 and M_4.

It should be noted that in this case, the output voltage is not equal to V_{DD} when the common mode voltage at the input is between V_{SS} and $V_{SS} + V_{TH}(M_1)$, due to the fact that charges are diode mounted PMOS transistors.

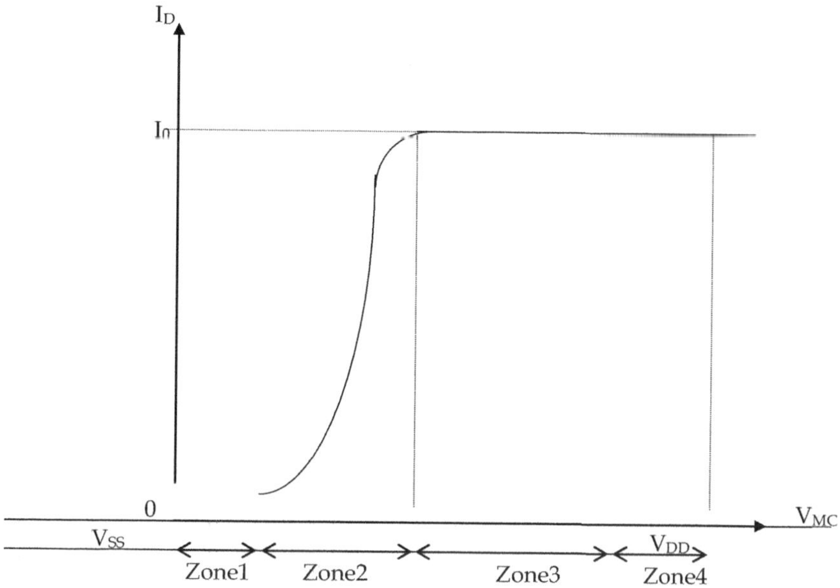

Figure 1.87. *Id (VMC) according to operating areas*

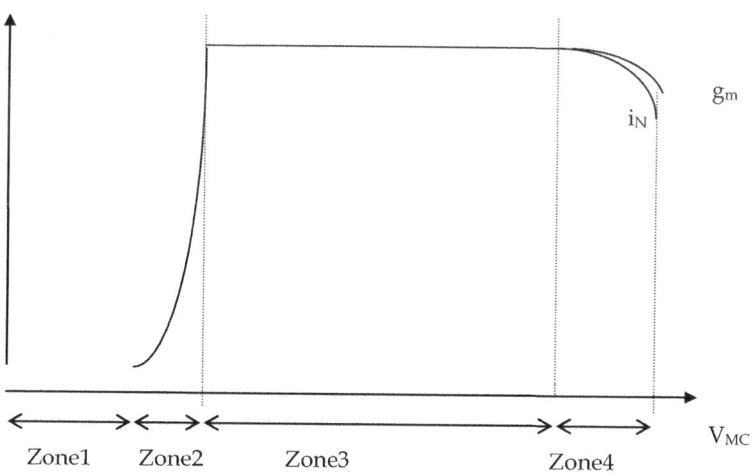

Figure 1.88. *I_N and g_m depending on the operating areas*

As a result, the nominal value of the low-frequency voltage gain is lower with "diode" transistors as charges, since the value of the equivalent resistance in small signals ($1/g_m$) is lower than the bias ratio V/I (R: DC resistance) (Figure 1.89).

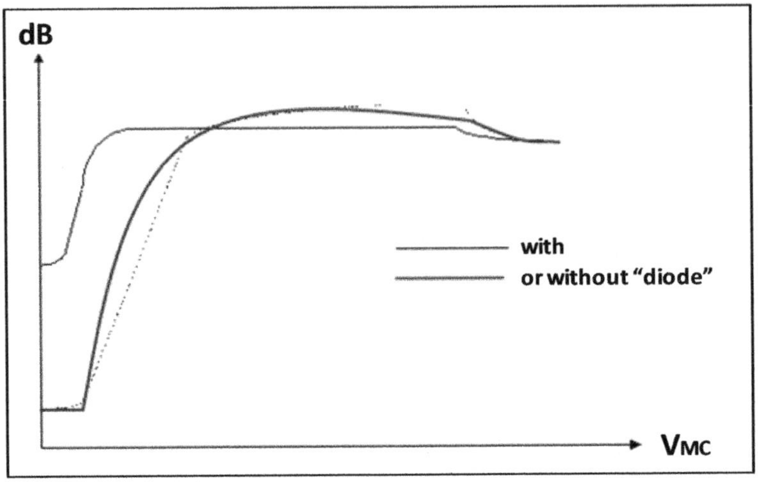

Figure 1.89. *Low-frequency voltage gain. For a color version of this figure, see www.iste.co.uk/gontrand/analog2.zip*

On the other hand, at low current, that is, for a common mode voltage close to V_{SS}, the transconductance g_m of M_3 and M_4 (without diode-connected transistors) becomes greater than the fixed value of resistors R_{Ni} (with diode-connected transistors), and the differential gain is then greater. Therefore, the loss of gain over the entire input dynamic range is less important.

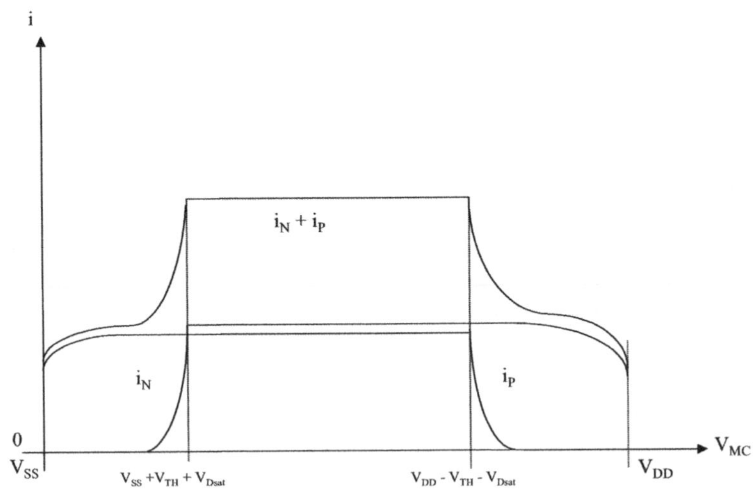

Figure 1.90. *i_N and i_P depending on common mode voltage*

The same study, applied to a PMOS differential stage, leads us to symmetrical results as the previous one in NMOS. This leads to the observation that there is a common mode voltage zone for which the amplitude of the current variation i_N (or i_P) cancelled out.

By referring to the same graph, these variations in current amplitude i_N and i_P as a function of the common mode voltage, it is interesting to note that when one of these amplitudes is cancelled out, the other is not cancelled out. Thus, the sum of i_N and I_P is never zero (Figure 1.90).

Analog design: MOS

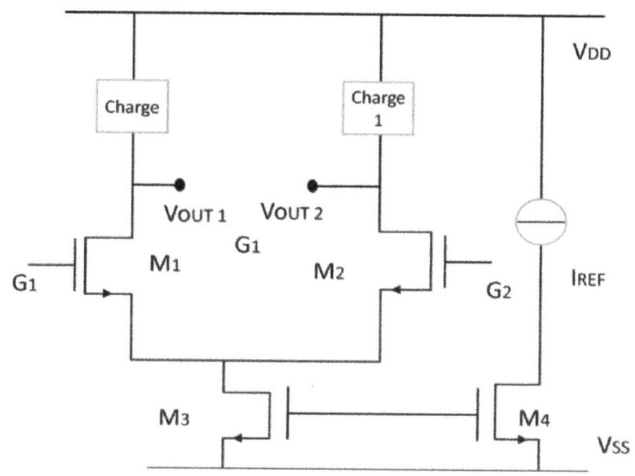

Figure 1.91. *NMOS differential pair. For a color version of this figure, see www.iste.co.uk/gontrand/analog2.zip*

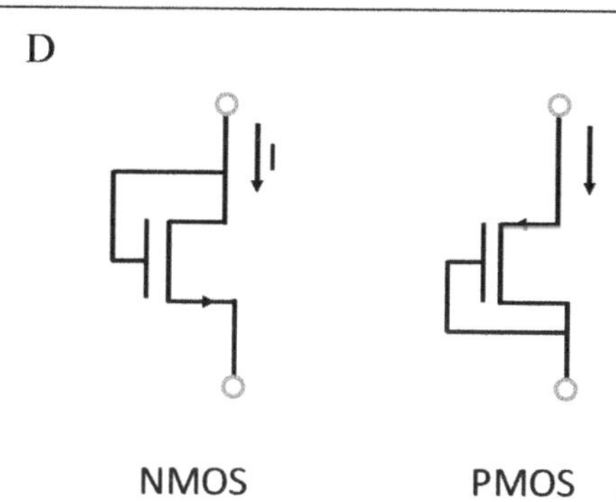

NMOS: I=-I_D

PMOS: I= +I_D

Note: The arrow is located at the source.
For the NMOS, it is outgoing.
For the PMOS: entering.

It indicates the direction of the current.

Figure 1.92. *Symbols for the two MOS. For a color version of this figure, see www.iste.co.uk/gontrand/analog2.zip*

In practice, a schematic is developed.

After defining the operation or quiescent point, this concerns the continuous, with a view to a small signal study (via SPICE; or, e.g. ELDO, ADS and HICUM).

Then: the design of the masks (geometry – layout) is realized.

It is a question, for example via Cadence (n.d.), of drawing the circuit, considering the rules of drawing (respecting minimum distances between transistors, e.g. design rule checker and layout versus schematic).

Finally, post-layout simulations are performed for verification.

Analog design of an inverter

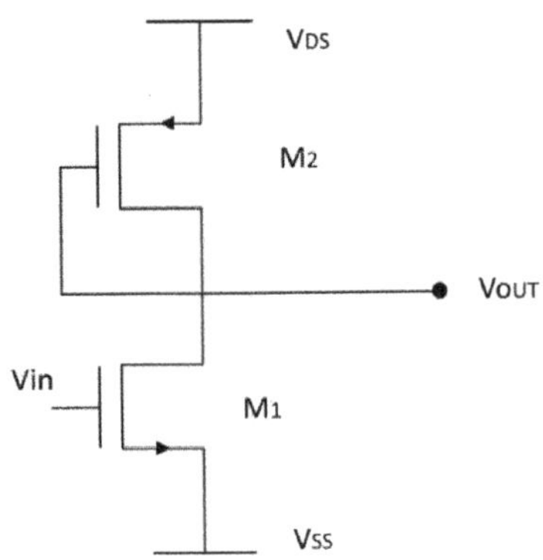

Figure 1.93. *NMOS inverter*

Vt_on = 0.8846 V

Kn = mn.Cox = 109 × 10^{-6} A/V^2

m_n = 487 cm^2/Vs

t_ox = 15.8 nm

Vt_op = –0.779 V

Kp = 44.1 × 10^{-6} A/V^2

μ_p = 197 cm^2/Vs

M_2

$V_{GS2} = V_{DS2}$

$V_{DS2} = V_{out} - V_{DD}$= 1-5= - 4V

$V_{top<0}$

$V_{DS2} < V_{GS2} - Vtop; so, M2\ is\ saturated$

Therefore:

$I_{DS2} = \frac{K_p}{2}\frac{W_2}{L_2}(V_{GS} - V_{top})^2$

M_1

$V_{in} = 3V$

$V_{GS1} = V_{in} - V_{SS}$= 3V

$V_{DS1} = V_{out} - V_{ton}$= 1-0= 1V; therefore: M_1: ohmic

$V_{GS1} - V_{ton}$= 3 – 0.884 ~ 2.10 V

$V_{ton} < V_{DS1} < V_{GS1}$; so: M_1: ohmic

$I_{DS1} = \frac{K_n}{}\frac{W_1}{L_1}(V_{GS1} - V_{ton} - V_{DS1}/2.).V_{DS1}$

$I_{DS1} = \frac{K_n}{2}\frac{W_1}{L_1}(2.(V_{GS1} - V_{ton}).V_{DS1} - {V_{DS1}}^2)$

M_2 dimensions

$|I_{DS1}| = |I_{DS2}|$

$$\frac{K_n}{2}\frac{W_1}{L_1}\left(2(V_{GS1} - V_{ton})V_{DS1} - {V_{DS1}}^2\right) = \frac{K_p}{2}\frac{W_2}{L_2}\left(V_{GS2} - V_{top}\right)^2$$

If $L_1 = L_2$:

$$\frac{W_2}{W_1} = \frac{K_n}{K_p}\frac{\left(2(V_{GS1} - V_{ton})V_{DS1} - {V_{DS1}}^2\right)}{\left(V_{GS2} - V_{top}\right)^2}$$

$V_{GS2} = V_{out} - V_{DD}$= - 2 - 5 = –7 V

$$\frac{W_2}{W_1} = \frac{109}{44.1} \frac{(2(3 - 0{,}884{,}6)1) - 1^2)}{(-4 + 0.779)^2}$$

$$\frac{W_2}{W_1} \sim 0.77$$

W1= 10 mm ➔ W2 (design) ~7.7 mm

Minimum V_{out}

When $V_{in} = 5V$

$V_{GS1} = V_{in} - V_{ss} = 5V$

$$I_{DS1} = \frac{K_n}{2}\frac{W_1}{L_1}(2(V_{GS1} - V_{ton})V_{DS1} - V_{DS1}{}^2)$$

$$I_{DS2} = -\frac{K_p}{2}\frac{W_2}{L_2}(V_{GS2} - V_{top})^2$$

$$I_{DS1} = -I_{DS2}$$

so:

$$K_n W_1(2(V_{GS1} - V_{ton})V_{DS1} - V_{DS1}{}^2) = K_p W_1(V_{GS2} - V_{top})^2$$

a.n. $V_{out} \sim 0.5$ V

Small signals

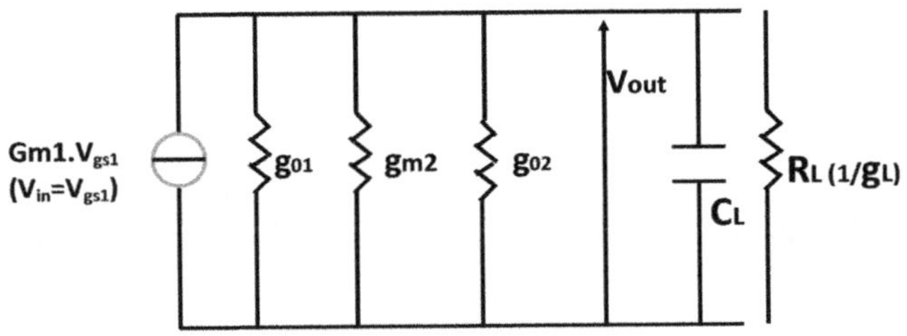

Figure 1.94. *Equivalent diagram in small signals. For a color version of this figure, see www.iste.co.uk/gontrand/analog2.zip*

Voltage gain (without capacitor)

$$\frac{V_{out}}{V_{in}} = \frac{V_{out}}{V_{GS1}} = -\frac{g_{m1}}{g_{01} + g_{m2} + g_{02} + (gL)} \quad [1.145]$$

Without load:

$$A_{v0}=-\frac{g_{m1}}{g_{01}+g_{m2}+g_{02}}$$

If $g_m \gg g_0$ $(saturated\ transistor)$:

$$A_{v0}\sim-\frac{g_{m1}}{g_{m2}}$$

$$g_m=\frac{\partial I_{DS}}{\partial V_{GS}}=\frac{K\,W}{L}(V_{GS}-V_t)$$

$$=\sqrt{2\frac{K\,W}{L}}\sqrt{\frac{K\,W}{L}(V_{GS}-V_t)^2}$$

$$A_{v0}=-\sqrt{\frac{K_n\left(\frac{W}{L}\right)_1}{K_p\left(\frac{W}{L}\right)_2}} \qquad [1.146]$$

a.n. A_{v0} ~ 1.9

Cut-off frequency:

The capacitor C_L is taken into account.

$$Z=-\frac{g_{m1}}{g_{01}+g_{m2}+g_{02}+\frac{1}{R_2}\left(1+j\frac{\omega}{\omega_C}\right)} \qquad [1.147]$$

where:

$$j\frac{\omega}{\omega_c}=1+j\omega\frac{C_L}{\Sigma\, g+\frac{1}{R_L}}$$

So:

$$\omega_c=\frac{g_{01}+g_{m2}+g_{02}+\frac{1}{R_L}}{C_L}$$

If $\frac{1}{R_L}\gg\Sigma\, g$

$f_c=\frac{1}{2\pi R_L R_C}$ ~ 16 kHz

f_c

Zin, Zout

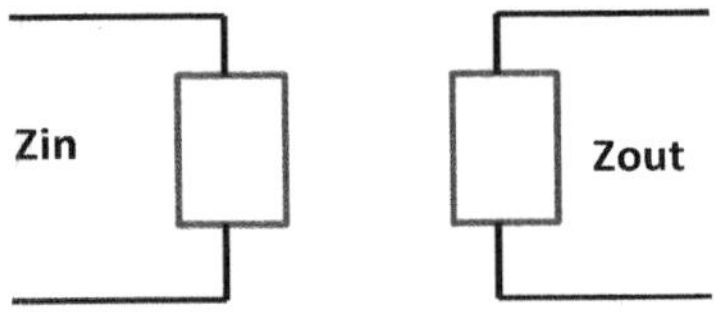

Figure 1.95. *Input and output impedance. For a color version of this figure, see www.iste.co.uk/gontrand/analog2.zip*

Zin=$\frac{1}{j\omega C_{GS}}$ (f decreases ➔ Zin increases).

Zout=$\frac{1}{g_{01}+g_{m2}+g_{02}+\frac{1}{R_2}}$

Cut-off frequency

We make a first-order assumption:

$g_m = \frac{1}{1+\tau p}$, with $\tau = X/\gamma_m \sim C_{GS}/g_m$; $p = j\omega$

$f_t \sim (1/2\pi).\ g_m/C_{GS}$

$f_t \sim 3/2.(\mu/2\pi L^2) = (V_{GS} - V_T)$

Small size = high f

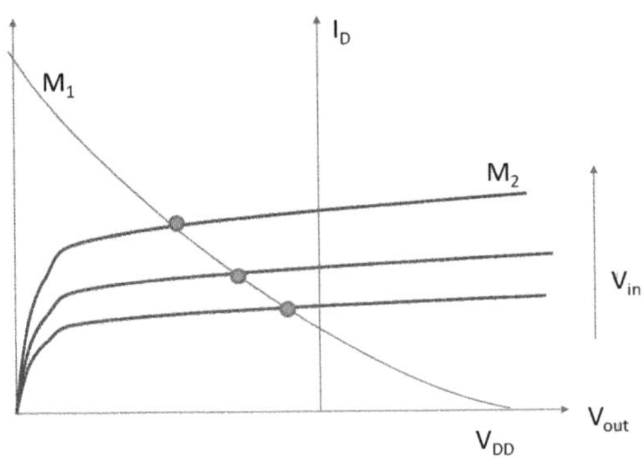

Figure 1.96. *I_D characteristics. For a color version of this figure, see www.iste.co.uk/gontrand/analog2.zip*

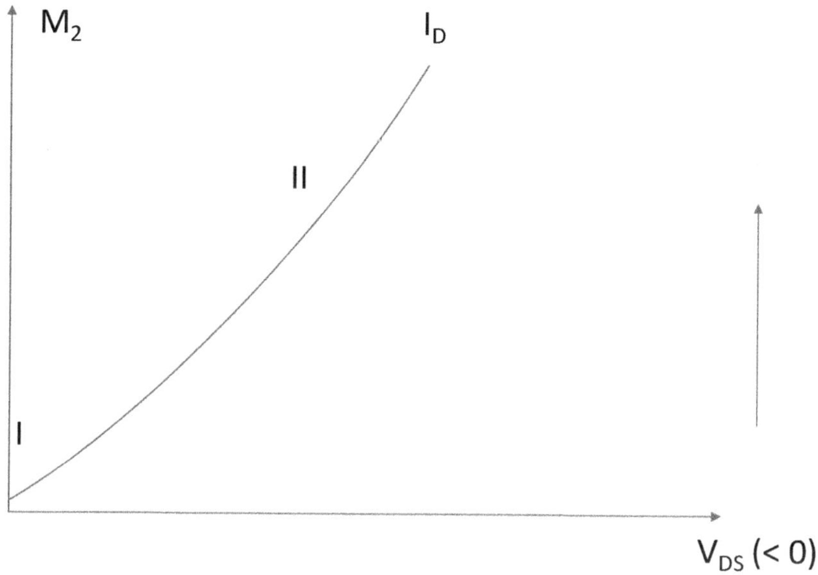

Figure 1.97. *I_D versus V_{DS}: saturated M_2. For a color version of this figure, see www.iste.co.uk/gontrand/analog2.zip*

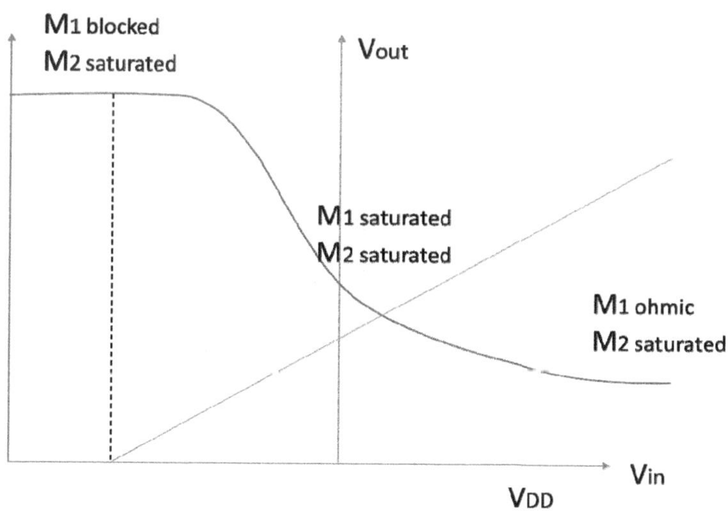

Figure 1.98. *Static regime for a NMOS inverter: output voltage versus input voltage. For a color version of this figure, see www.iste.co.uk/gontrand/analog2.zip*

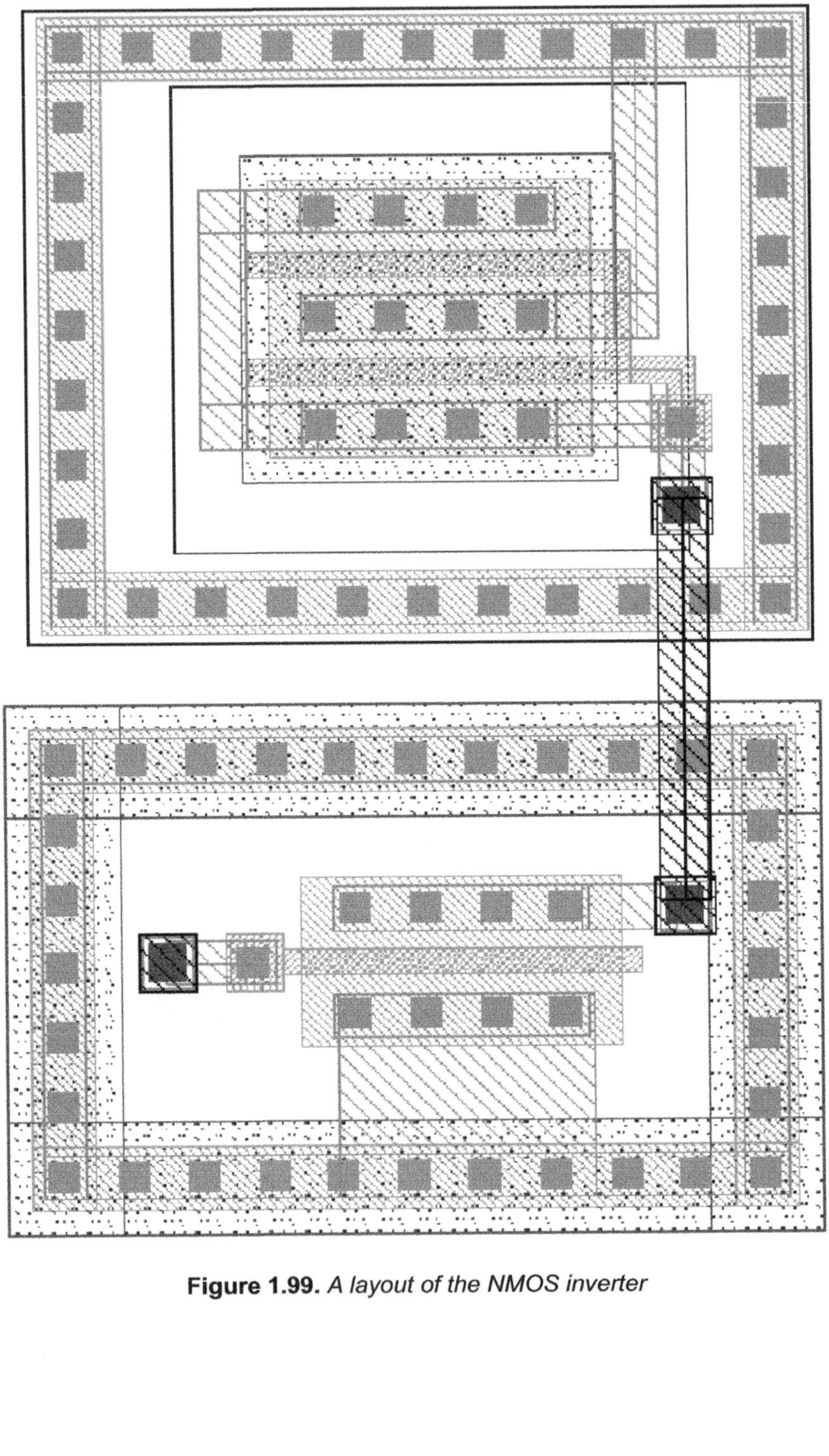

Figure 1.99. *A layout of the NMOS inverter*

Differential pair

Active charge: PMOS

When transistors are in saturation:

$$I = \frac{K}{2}\frac{W}{L}\,(V_{GS} - V_T)^2$$

or:

$$I = \frac{\beta}{2}\,(V_{GS} - V_T)^2 \qquad [1.148]$$

With:

$$\beta = K\frac{W}{L}$$

$$V_{GS} = V_{DS} = V_T - \sqrt{\frac{2I}{\beta}}$$

Preliminary calculations

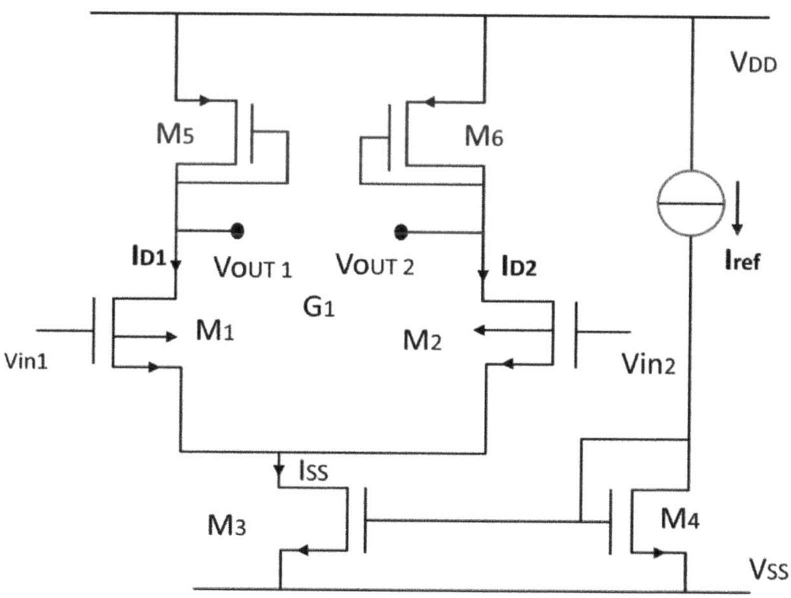

Figure 1.100. *MOS differential amplifier. For a color version of this figure, see www.iste.co.uk/gontrand/analog2.zip*

$$V_{D1} = V_{DD} + V_{DS5}$$

$$V_{D2} = V_{DD} + V_{DS6}$$

$$V_{D1} = V_{DD} + V_{T5} - \sqrt{2\frac{I_{D2}}{\beta_5}} \quad [1.149]$$

$$V_{D2} = V_{DD} + V_{T6} - \sqrt{2\frac{I_{D2}}{\beta_6}} \quad [1.150]$$

We also have:

$$I_{D1} = \frac{I_{SS}}{2} + \frac{I_{SS}}{2}\left(\beta\frac{V_{SS}{}^2}{I_{SS}} - \frac{\beta^2 V_{ID}{}^4}{4I_{SS}{}^2}\right)^{1/2}$$

$$I_{D2} = \frac{I_{SS}}{2} - \frac{I_{SS}}{2}\left(\beta\frac{V_{SS}{}^2}{I_{SS}} - \frac{\beta^2 V_{ID}{}^4}{4I_{SS}{}^2}\right)^{\frac{1}{2}} \quad [1.151]$$

Voltage gain:

$$A_V = \left.\frac{\partial V_{D1}}{\partial V_{ID}}\right|_{V_{ID} \sim 0}$$

$$A_V = \left.\frac{\partial V_{D1}}{\partial I_{ID1}} \cdot \frac{\partial I_{ID1}}{\partial V_{ID}}\right|_{V_{ID} \sim 0}$$

$$V_{D1} = V_{DD} + V_{T5} - \sqrt{2\frac{I_{D1}}{\beta_5}}$$

So:

$$\frac{\partial V_{D1}}{\partial I_{ID1}} = -\sqrt{\frac{I_{D1}}{2\beta_5}}\sqrt{\frac{1}{2I_{D1}}}$$

But $V_{ID} \sim 0$; $so\ I_{ID1} \sim \frac{I_{SS}}{2}$

$$A_V = -\frac{1}{2}\sqrt{\frac{2}{\beta_5}}\sqrt{\frac{2}{I_{SS}}}\sqrt{\beta_1\frac{I_{SS}}{4}}$$

$$A_V = -\frac{1}{2}\sqrt{\frac{\beta_1}{\beta_5}}$$ with $\beta = K.\Omega/\Lambda = \mu\ X_0.W/L$

$$A_V = -\frac{1}{2}\left(\frac{\mu_n C_O \left(\frac{W}{L}\right)1}{\mu_n C_O \left(\frac{W}{L}\right)5}\right)^{1/2}$$

where $\beta = KW/L = \mu.CoxW/L$:

$$A_V = -\frac{1}{2}\left(\frac{\mu_n C_O}{\mu_n C_O}\frac{W_1 L_5}{W_5 L_1}\right)^{1/2}$$

$W_1 = 10\ \mu\mu = \Omega_2$

$L_1 = 1.2\ \mu\mu = \Lambda_2$

$W_5 = 10\ \mu\mu = \Omega_6$

$L_5 = ?\quad L_6 = ?$

Determination of L_5 and of L_6 to get a gain of 3

$$A_V = -\frac{1}{2}\left(\frac{\mu_n}{\mu_p}\frac{W_1}{W_5}\frac{L_5}{L_1}\right)^{1/2}$$

Therefore:

$$L_5 = 4|A_V|^2\frac{\mu_p}{\mu_n}L_1 \sim 11.5\ \mu\mu$$

We can perform simulations, for comparisons, via the "schematic" and the "layout" under the Cadence (n.d.) standard, for example.

Voltage multiplier circuits

Two quadrant multiplier

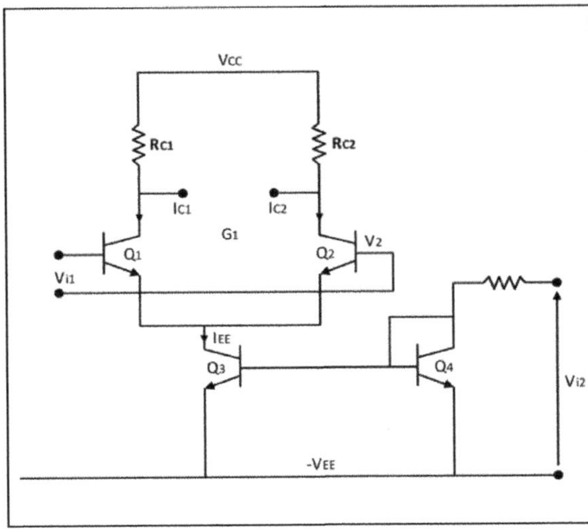

$$I_{C1} = \frac{I_{EE}}{1 + exp\,(-\frac{V_{i1}}{V_T})}$$

$$I_{C2} = \frac{I_{EE}}{1 + exp\,(\frac{V_{i1}}{V_T})}$$

$$\Delta I_C = I_{C1} - I_{C2} = I_{EE}\,tanh\frac{V_{i1}}{2V_T}$$

$$\frac{V_{i1}}{2V_T} \ll 1 \quad tanh\frac{V_{i1}}{2V_T} \sim \frac{V_{i1}}{2V_T}$$

$$I_{EE} = \frac{V_{i2} - V_{BE3}}{R}$$

$$\Delta I_C = \frac{1}{2R.V_T} V_{i1}(V_{i2} - V_{BE3})$$

$$V_{i2} \gg V_{BE3}$$

Figure 1.101. *Two quadrant multipliers*

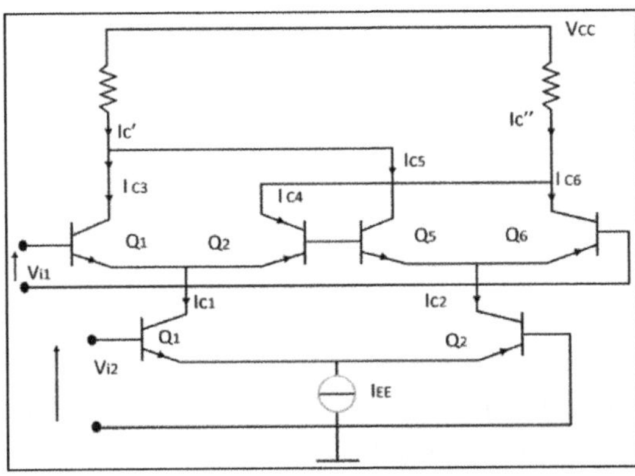

$$I_{C3} = \frac{I_{EE}}{1 + exp\,(-\frac{V_{i1}}{V_T})}$$

$$I_{C4} = \frac{I_{EE}}{1 + exp\,(\frac{V_{i1}}{V_T})}$$

$$\Delta I_C = I'_C - I''_C =$$

$$\Delta I_C = I_{EE}\,tanh\frac{V_{i1}}{V_T} . tanh\frac{V_{i1}}{V_T}$$

Figure 1.102. *Analog multiplier (differential input, one output). For a color version of this figure, see www.iste.co.uk/gontrand/analog2.zip*

Analog multiplier

$$\Delta I_C = I'_C - I''_{C=}$$

$$\Delta I_C = I_{EE}\frac{K_1 V_{i1}}{I_{01}}.\frac{K_2 V_{i2}}{I_{01}}$$

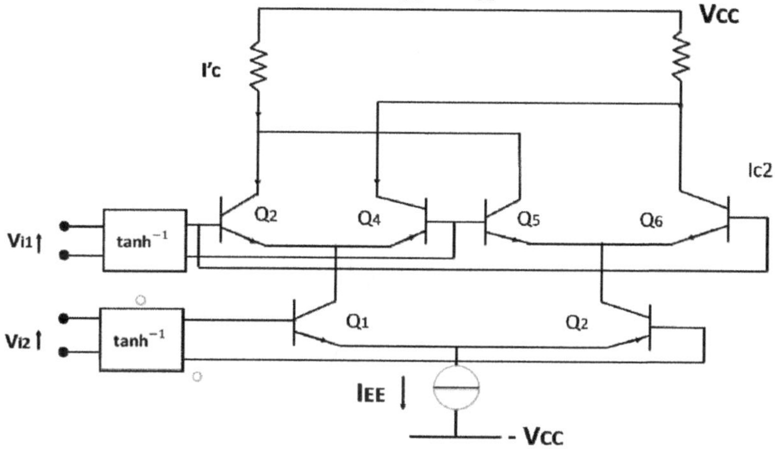

Figure 1.103. *Analog multiplier. For a color version of this figure, see www.iste.co.uk/gontrand/analog2.zip*

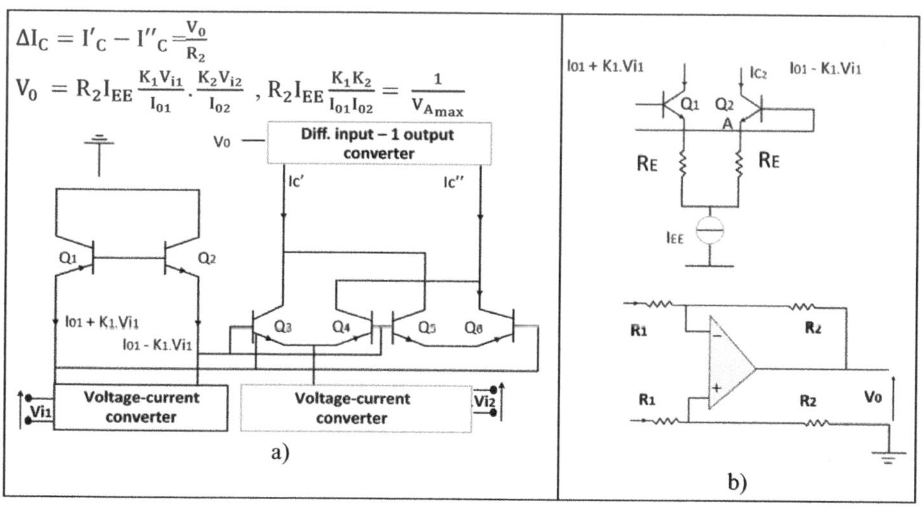

Figure 1.104. *Two inputs/one output converter. For a color version of this figure, see www.iste.co.uk/gontrand/analog2.zip*

Phase locked loop (PLL) systems

Phase detector

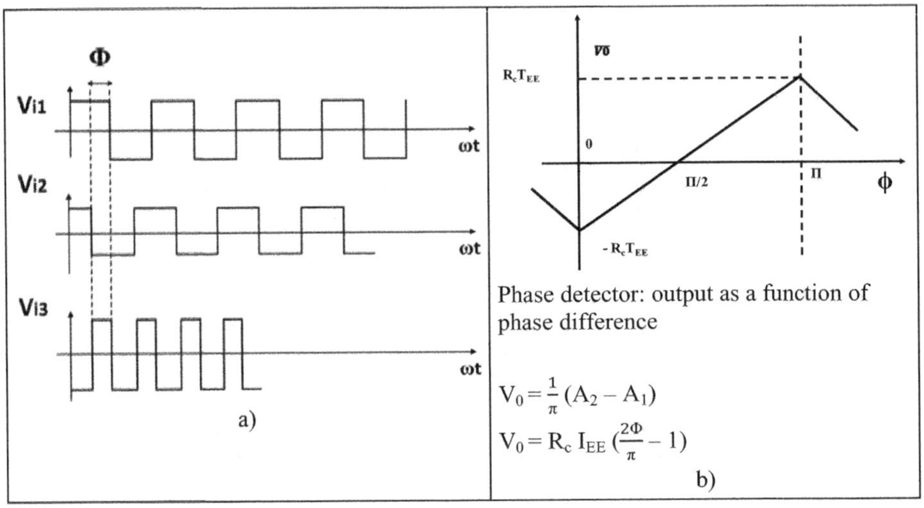

Figure 1.105. *Phase detection*

Below is a block diagram of a phase locked loop (PLL).

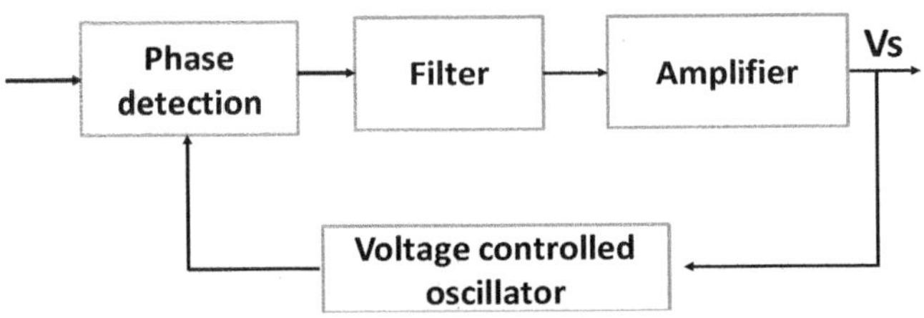

Figure 1.106. *Block diagram of a PLL. For a color version of this figure, see www.iste.co.uk/gontrand/analog2.zip*

The phase detector produces a DC or LF voltage proportional to the phase difference between the incoming signal and the VCO output signal. The signal related to the phase difference then passes through a loop filter and then an amplifier and is then applied to the VCO control input.

If the frequency of the input signal varies slightly, the phase difference between the VCO signal and the input signal will increase as a function of time.

This will change the control voltage on the VCO so as to bring the frequency of the VCO back to that of the input signal.

Suppose that the frequency results in that of the freewheeling frequency. The frequency of the frequency difference wave is reduced when its incremental amplitude is negative, and increases when the amplitude is positive: then the output of the phase detector exhibits an asymmetry in the waveform of the signal, during the capture (Figure 1.104), the latter introducing a continuous device at the output of the detector, which shifts the mean frequency from the VCO to the frequency of the input signal and gradually decreasing the frequency difference. Since the frequency difference is cancelled out, it remains a continuous device at the output of the filter.

The loop capture range is the range of input frequencies around the center frequency on which the loop will be locked from a lock condition.

Reducing the filter bandwidth improves rejection of out-of-band signals but will reduce the capture range.

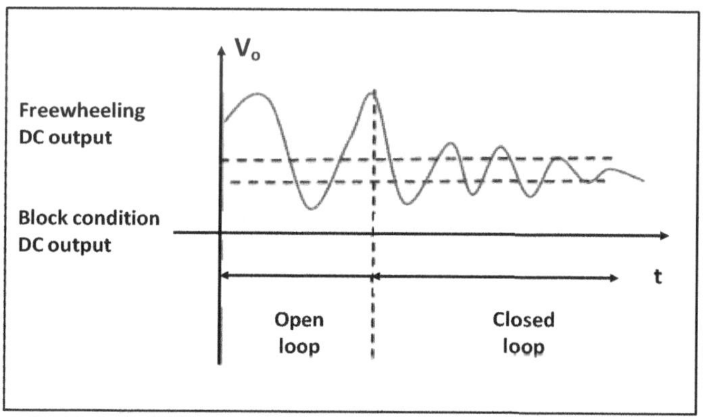

Figure 1.107. *Detector output. For a color version of this figure, see www.iste.co.uk/gontrand/analog2.zip*

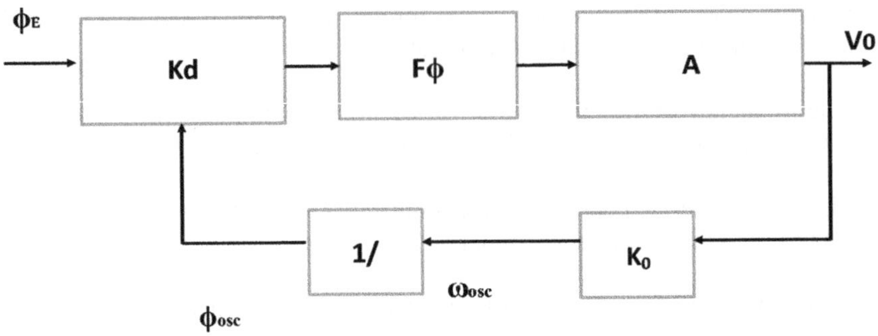

Figure 1.108. *Block diagram of a PLL. For a color version of this figure, see www.iste.co.uk/gontrand/analog2.zip*

$$\omega_{osc} = \frac{d\Phi osc\ (t)}{dt}$$

Let us integrate:

$$\Phi osc = \Phi osc(t = 0) + \int_0^t \omega_{osc}\,(t)dt$$

The relationship between the output frequency of the VCO and V_o can be written as follows:

$$\omega_{osc} = \omega + K_O V_O$$

Therefore, the feedback theory can be applied to the system, which is linear.

$$\frac{V0}{\Phi i} = \frac{KdF(p)A}{1 + KdF(p)AKo/p}$$

Starting from:

$$\omega i = \frac{d\Phi i\ (t)}{dt}$$

and:

$\omega_o\,(p) = p.\Phi i(p)$, where p is a complex number

$$\frac{V0}{\omega i} = \frac{KdF(p)A}{p + KdF(p)AKo}$$

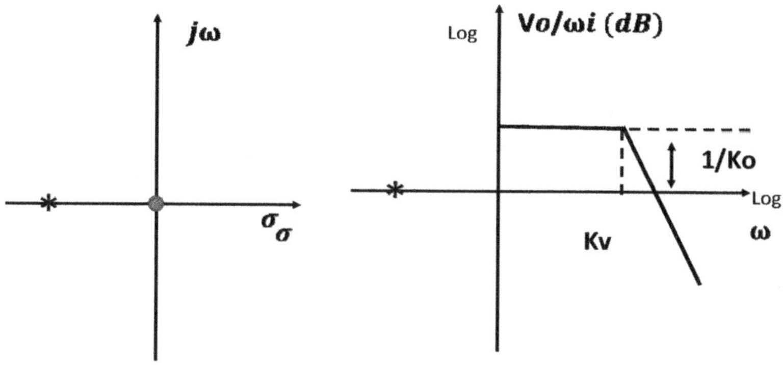

Figure 1.109. *Pole/zero loci and voltage gain (in dB). For a color version of this figure, see www.iste.co.uk/gontrand/analog2.zip*

a) If F(p) = 1, we end up with a low-pass filter of the first order (an integrator).

$\frac{V_0}{\omega i} = \frac{Kv/K_0}{p+Kv}$, with: $Kv = K_0 K_d A$: loop width, for the modulating signal, demodulated by the PLL.

$$F(\pi) = \frac{1}{1+\frac{p}{\omega_1}}$$

$$\frac{V_0}{\omega_i} = \frac{1}{K_0}\frac{1}{1+\frac{p}{\omega_1}+\frac{p^2}{Kv}}$$

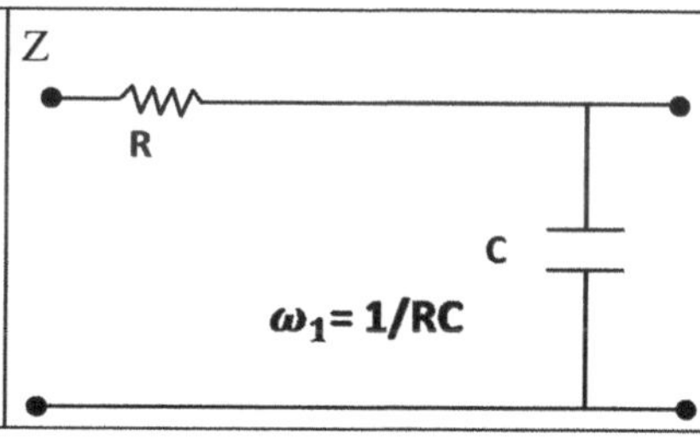

Figure 1.110. *Transfer function*

b) $F(p) = \frac{1}{1+p/\omega 1}$

$$\frac{V_0}{\omega i} = \frac{1}{K_0}\frac{1}{1+\frac{p}{Kv}+p2/\omega i Kv}$$

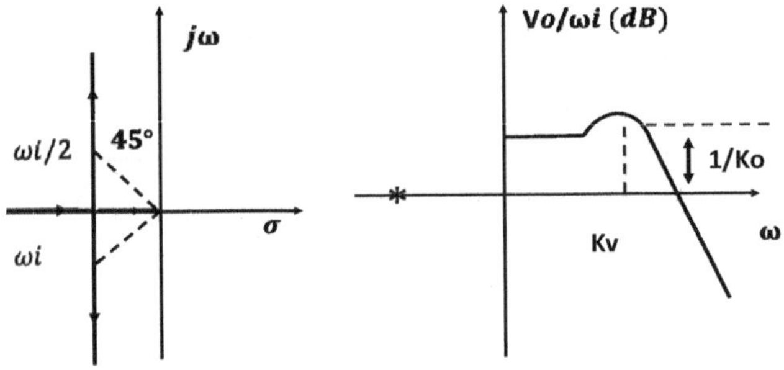

Figure 1.111. *Root loci and voltage gain transfer function*

Transfer function:

$$\frac{V_O}{\omega i} = \frac{1}{K_O}\frac{1}{1+\frac{2T}{\omega_n}p+p^2/\omega_1^2}$$

with:

$$\omega_n = \sqrt{K_1\omega_1} \quad T = \ 1/2\sqrt{\frac{\omega_1}{K_v}}$$

$$T = 0.7\ \omega_1 = 2\ Kv\ w_n = 1.4\ K_v\ (-3\ dB)$$

1) Phase detector

– Gilbert multiplier

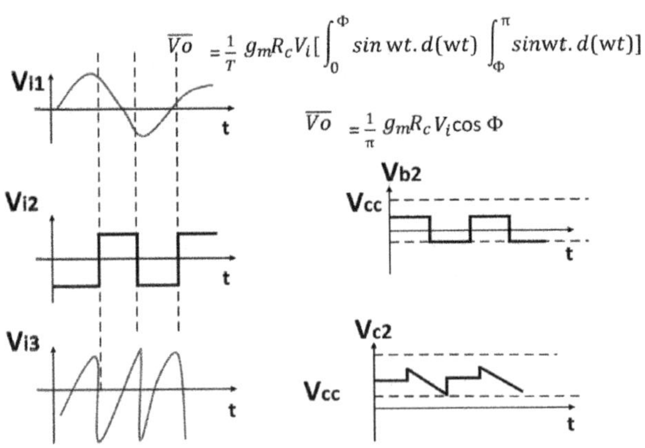

Figure 1.112. *Phase detector wave multiplied by a synchronous square wave signal. For a color version of this figure, see www.iste.co.uk/gontrand/analog2.zip*

2) Voltage controlled oscillator

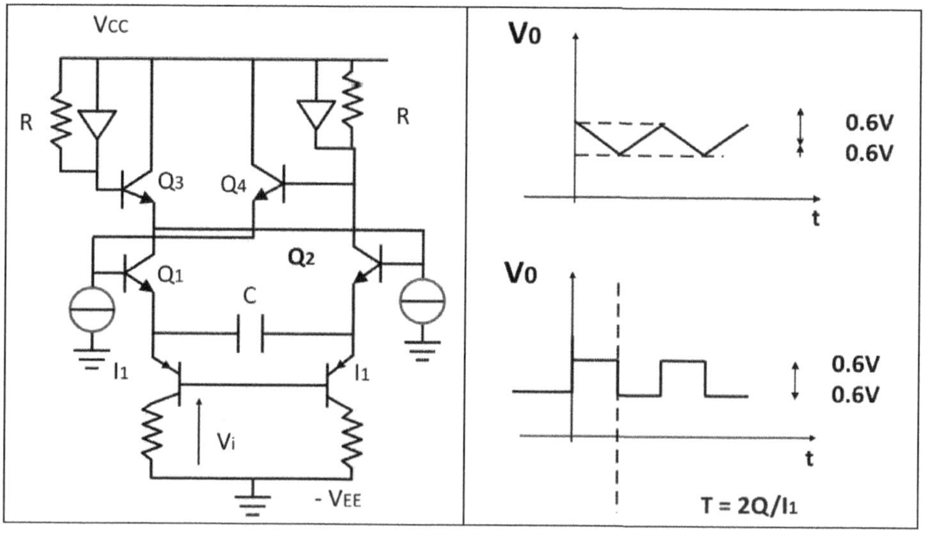

Figure 1.113. *A VCO. For a color version of this figure, see www.iste.co.uk/gontrand/analog2.zip*

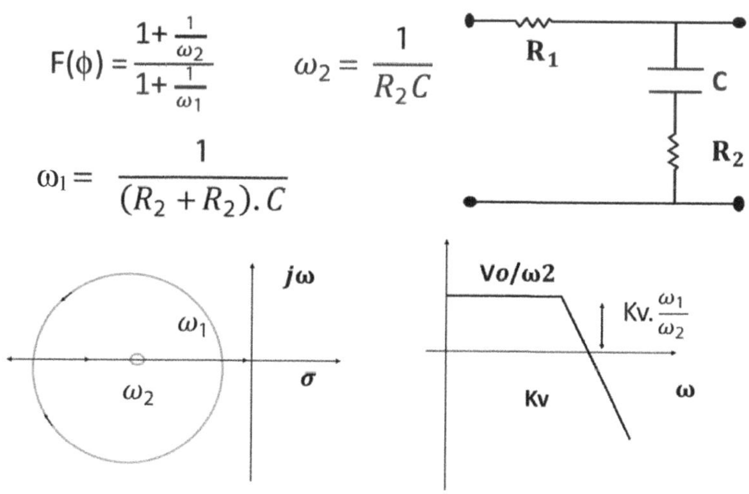

Figure 1.114. *Root loci and voltage gain transfer function. For a color version of this figure, see www.iste.co.uk/gontrand/analog2.zip*

Lock range

$$K_D p/2 \geq V\ comp\_phase \geq -K_D p/2$$

$$\Delta w_{osc} = K_D A K_0 \pi/2 = K_v \pi/2 = \Delta w_L/2$$

——*——————*——————————*——

Latching loss ——<——————>—*<——————> Latching loss,

$\Delta w_L/2$ $\Delta w_L/2$

Capture range

$$\Delta w_c = \Delta w_L$$

$$\Delta w_c = \Delta w_L .\ |F(\Delta w_c)|$$

1.9. Conclusion

We presented an introduction to analog circuits via modeling for simulations, such as nodal analysis.

This b-a-ba should allow an initiation for some, knowing the supremacy of the digital aspect, which is certainly unavoidable. This is a structural problem. Analog is the acquisition, in terms of concepts at first, not so many as that (although we did not talk about radio or microwave frequencies, where the approach is more physical), of know-how in terms of experience: essential.

The industry seems to be struggling to find well (self-)trained young people in this exciting field, in this case at the college diploma and master's levels, which, in particular, leads to more than attractive opportunities.

2

Noise and Interference in Mixed Circuits

This chapter investigates substrate coupling effects in mixed integrated circuits, particularly perturbations on radiofrequency (RF) blocks. The design and analysis of fully integrated voltage-controlled oscillators (VCOs) is one of the key points of RF analysis. First, sensitivity functions at the oscillation frequency of the control voltage, the bias current and parasitic sidebands due to the injected noise are extracted to discover a relationship between the noise of the substrate and the purity of the spectrum.

At the heart of the section, we try to develop quantitative predictions on the phase noise of these oscillators and give new leads in this field. Mixed-mode simulations are involved by applying a microscopic drift scattering model, while Kirchhoff's laws govern the rest of the circuit used. Another problem for designers of complex heterogeneous circuits is predicting perturbations coming from switched logic gate blocks, flowing through the substrate to reach certain sensitive analog blocks. We present an application of a stochastic process model; the activity of digital switching is managed via Markov chains.

2.1. Introduction

More and more system-on-chip designs require the integration of very fast analog circuits and large digital blocks. Until now, perturbations introduced by the digital part with respect to the analog part are a key problem and difficult to manage by designers, in particular, in the RF range. Pragmatically, the first objective of this chapter is to simply present the problem and propose some basic rules associated with a methodology and some tools to quantify and reduce the noise of the substrate in the future design, implied by the digital part, which will degrade the performance of analog circuits (RF). The problem in depth is to build efficient models for the activity of the digital part, the substrate and their coupling, in a package that could be integrated into a single computer-aided design (CAD) tool.

An important microwave function in transceivers is the generation of millimeter waves. However, there is no generic method for designing a low-phase noise oscillator. Usually, people have their homemade "know-how", and the low-phase noise characteristic is obtained only by using a prototype that is difficult to reproduce for industrial implementation. Moreover, oscillator characteristics, such as power consumption, output power, and adjustment range, are very important and are also optimized. It is therefore very important to develop a generic method, which will make it possible to obtain a good precision of integrated circuit design.

2.2. Ground or power supply noise and substrate coupling

From our point of view, substrate perturbations could be the central subject when dealing with very complex circuits. We will describe the different injection mechanisms in a digital circuit. Figure 2.1 shows a CMOS (complementary MOS) inverter and the various sources of its substrate noise.

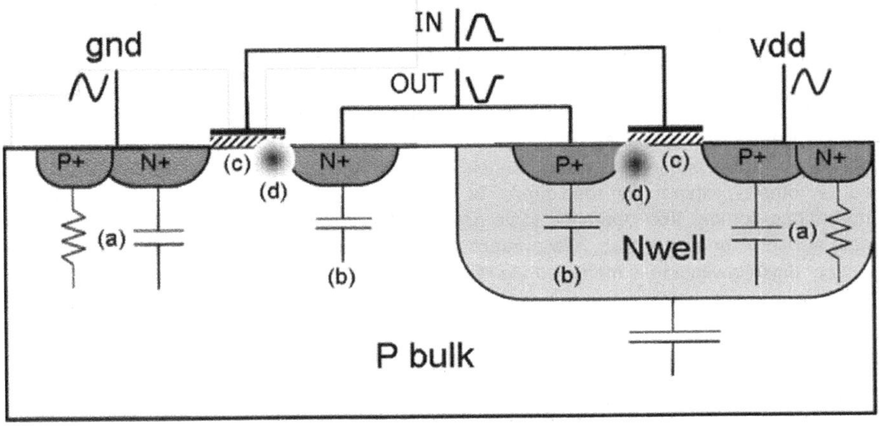

Figure 2.1. *Mechanisms for injecting substrate noise into a digital inverter*

The main source of substrate noise is due to the power supply and ground bounces that are transmitted through the substrate via polarized substrate contacts; in a CMOS inverter, the source of the hot pole. Power and mass fluctuations are generated by large spikes in transient currents drawn by the simultaneous switching of several digital gates. These peaks, as well as the parasitic elements of the packaging, the PCB (Printed Circuit Board), the padring and other parts of the supply and ground networks, can introduce resonance channels and therefore a power and ground resonance. Supply oscillations are directly coupled to the substrate by the P+ polarization viaticas, as shown in Figure 2.1. This phenomenon is the main source of substrate noise in mixed-mode integrated circuits that propagates in the substrate via resistive contacts.

Other mechanisms that inject parasitic currents into the ground have capacitive behavior. First, the local desertion capacitance at the drain (b) of each MOS transmits variations in the substrate. In the case of an inverter, the output voltage is directly coupled to the substrate because of these "depletion" capacities of the NMOS and PMOS drains. In addition, the gate electrode is coupled to ground through the gate oxide (c) and the channel capacitance. The input voltage of the gate is also transmitted to ground. The signals can also be transmitted in the substrate through all the metal interconnections of the chip. The capacitance between the metal line and the substrate strongly depends on the length of the interconnection.

Thus, the contribution of the latter noise source is highly dependent on the design and can be reduced by reducing the length of the interconnection lines and then the coupling of capacitance between the signal and ground. Another noise generator is the well-known impact ionization (d). When an MOS transistor is biased into the saturation state, a high electric field develops in the depleted region of the channel near the drain. This high field generates hot carriers that can dissipate their energy by creating electron-hole pairs. For an NMOS, the holes created are scanned in the substrate. Other minor noise sources are described in various developments such as gate-induced drain leakage, photon-induced current, and diode leakage current. All carriers created by the injectors are propagated through the substrate.

These perturbations affect analog blocks in the same way as they were generated. The coupling with the substrate may be resistive through bias contacts of the substrate, capacitive through interconnections and depletion capacitances, nonlinear through the MOS body effect mechanism for example or other effects related to each basic analog device. Variations in the operation of each device can cause serious signal integrity problems in the overall analog system, which can significantly decrease its own performance.

In this section, we show a simple model to simulate substrate tension. We will consider power and ground bounces as the sole source of substrate noise to significantly simplify the problem and simulation models. Considering only a substrate layout model used to supply contacts and wells, makes it possible to analyze, without significant inaccuracy, the noise of the substrate generated by digital blocks. Power and ground lines must be well modeled to predict voltage drop. Thanks to the ICEM (Integrated Circuit Emission Model), emission levels in conducted mode can be predicted on a chip as well as in the application card by means of analog simulations based on SPICE. Several papers show a good correlation between measurements and simulations. Thus, this model is suitable to predict power and ground bounces, and therefore the tension of the substrate.

2.2.1. *Noise propagation in a silicon substrate*

After the substrate noise generation described in the previous section, propagation must be quantized through the substrate. The silicon material is characterized by a cutting frequency $f_c = \sigma/_{2\pi\varepsilon}$, where σ is the conductivity and ε the permittivity. With a high substrate resistivity (>10 Ω·cm), silicon can be considered as purely ohmic for the signal frequency below 10 GHz. Silicon can be considered as purely ohmic for a signal frequency below 10 GHz. Then, the equivalent transfer function of the silicon can be defined as an attenuation factor that can be calculated with an equivalent point-to-point resistance. At this step, it is necessary to take into account the nature of the substrate: low- or high-resistivity silicon substrate. Given the substrate with low resistance, the substrate model depends on the comparison between the thickness of the epitaxial layer (W_{epi}) and the distance D between substrate contacts. As shown in Figures 2.2 and 2.3 (case b), when the distance is above all current flow, lines cross the low-resistance substrate, which can be considered as a single node. The distance resistance variation function is represented in Figure 2.4 (the substrate contact zone is 20 × 20 μm², W_{epi} = 5 μm at 10 Ω·cm; the thickness of the substrate is 300 μm × 0.05 Ω·cm).

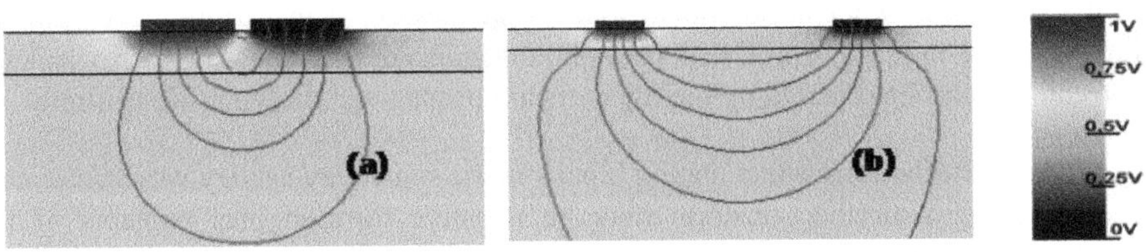

Figure 2.2. *Current flow lines for low-resistance substrate layer, with epitaxial layer (thickness): (a) $D < 4 \times W_{epi}$; (b) $D > 4 \times W_{epi}$. For a color version of this figure, see www.iste.co.uk/gontrand/analog2.zip*

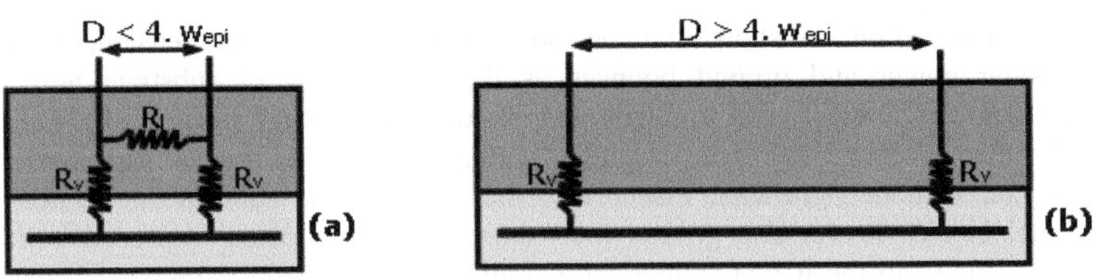

Figure 2.3. *Equivalent model for low-resistance substrate with epitaxial layer (thickness): (a) $D < 4\times W_{epi}$; (b) $D > 4\times W_{epi}$*

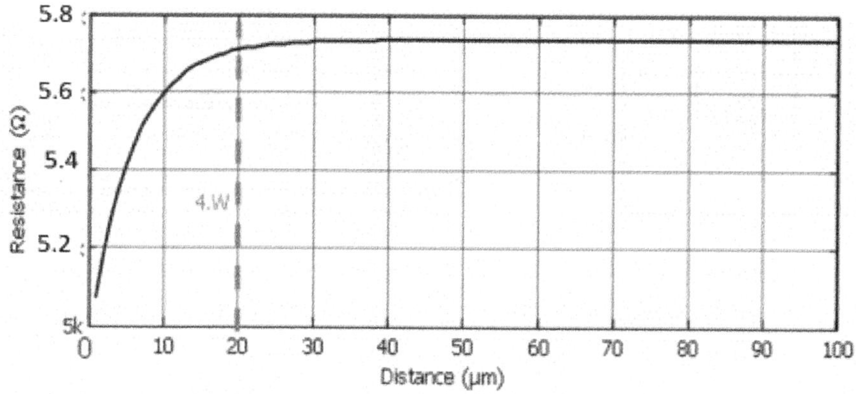

Figure 2.4. *Distance equivalent resistance function for low-resistance substrate with epitaxial layer (thickness)*

For a high-resistance substrate, the back connection of the matrix is of primary importance. If the back of the "bulk" floats, resistance between two contacts increases with distance. If the back of the matrix is properly grounded, resistance depends on the distance between contacts relative to the thickness of the substrate $W_{substrate}$, as illustrated in Figures 2.5–2.7 (substrate contact area is 20 × 20 µm², $W_{substrate}$ = 50 µm at 6.7 Ω·cm).

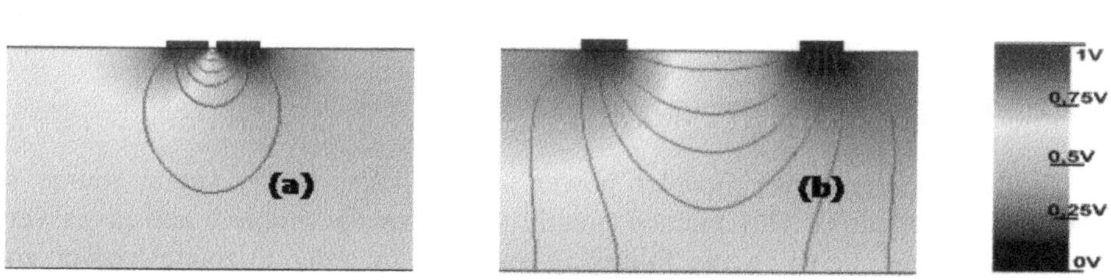

Figure 2.5. *Current flow lines for high-resistance substrate (thickness) with grounded back: (a) $D < W_{substrate}/2$; (b) $D > W_{substrate}/2$. For a color version of this figure, see www.iste.co.uk/gontrand/analog2.zip*

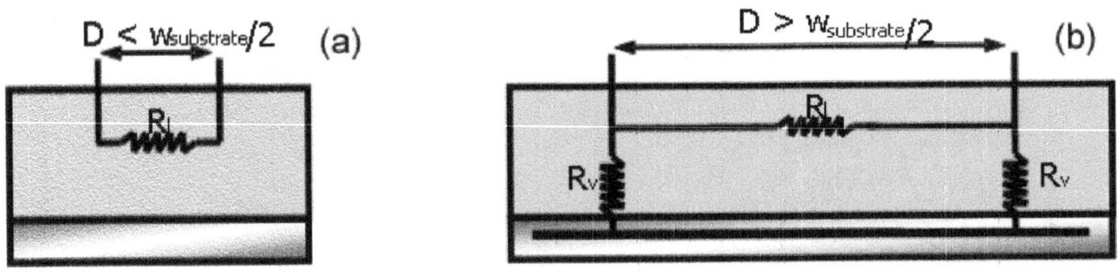

Figure 2.6. *Equivalent model for high-resistance substrate (thickness) with grounded back: (a)* $D < {W_{substrate}}/{2}$ *; (b)* $D > {W_{substrate}}/{2}$

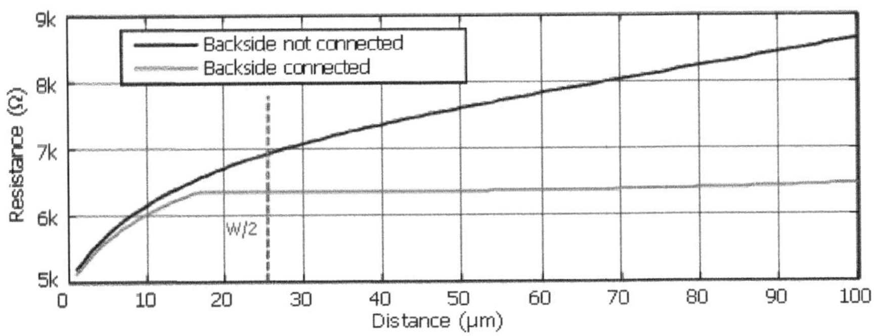

Figure 2.7. *Distance equivalent resistance function for high-resistance substrate (thickness) with floating or grounded back*

2.2.2. *Simulation methodology*

In this approach, we consider power supply noise as the single source of substrate perturbation. This method greatly simplifies the problem and gives very good results, as shown in the following sections. In order to model power and ground voltage oscillations, we therefore chose to use the ICEM (Integrated Circuit Electromagnetic Model) approach. Classic ICEM has been extended with a substrate array sub-model. Figure 2.8 provides an overview of the extended ICEM model.

The passive distribution network sub-model is expected to model parasitic elements of the power supply or signal lines. The internal activity sub-model characterizes the dynamic current consumption of the system. The substrate-network sub-model should model the propagation in the substrate of parasitic signals from the aggressor parties to the victim parties.

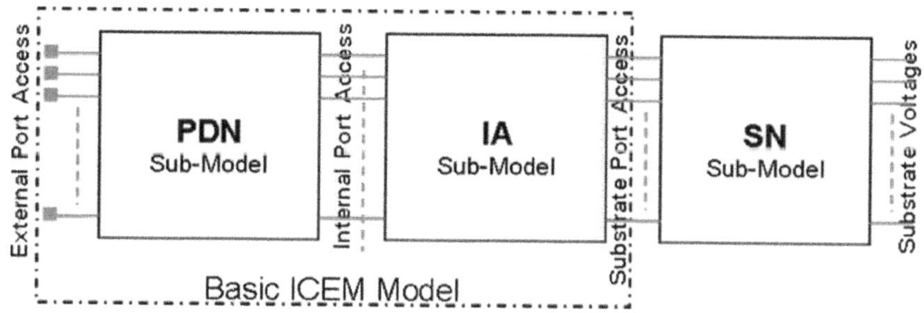

Figure 2.8. *Conceptual architecture of the extended ICEM model*

Tools developed

Basically, the extended ICEM conceptual methodology can be described in its simplest form by the electrical diagram shown in Figure 2.9 (a digital block with a V_{DD}-$_{GND}$ power pair). Nevertheless, certain parameters must be determined for each IC project. Therefore, we propose using two software applications developed in JAVA (Figure 2.10). The first application is developed for the bounce effects of mass and power. The second concerns the extraction of the substrate. These applications can be used at any stage of the design phase: from the preliminary study to the final optimization (post-page layout), because parameters can be refined over time.

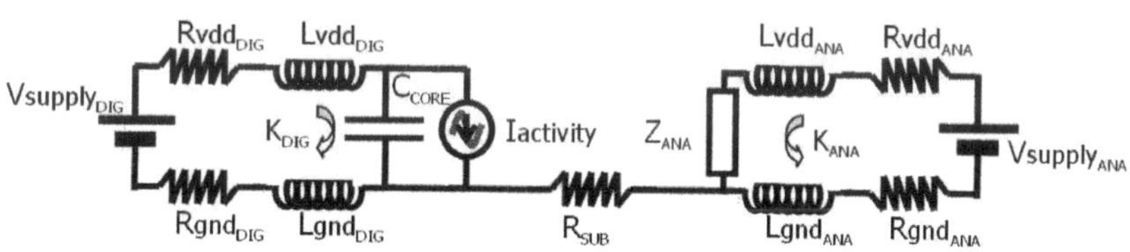

Figure 2.9. *Basic architecture of the extended ICEM model. For a color version of this figure, see www.iste.co.uk/gontrand/analog2.zip*

Applications developed for ground and power bounce effects

The first application deals with the quantification of fluctuations in mass and power supply (bounces). This can be done in three steps. First, the user describes passive elements: R, L, C_{core} and K. Parameters R and L are the parasitic resistance and the inductance of the lines (PCB, socket, housing, metal rails on chip). K is the coupling factor (mutual inductance) between lines (mutual between bonding wires, the typical value is about 0.5). In digital integrated circuits, parasitic power and ground elements are mainly due to the effects of packaging; the effects of the power grid can be overlooked in a first-rate approach. Packaging lead and bonding wire lengths are often longer than metal lines on silicon and mainly induce parasitic inductances. C_{core} is the equivalent numerical capacity; this depends on the technology (number of gates per mm²) and the numerical area. Typical values are 0.15–0.3 nF/mm² for standard 0.35 µm CMOS technology and 0.8–1.6 nF/mm² for 90 nm. These passive parasitic elements can be estimated or determined from the measurement/parameters S.

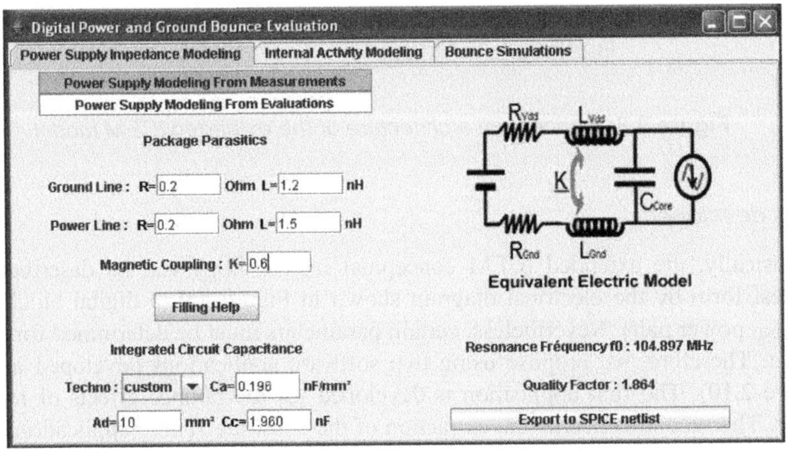

a)

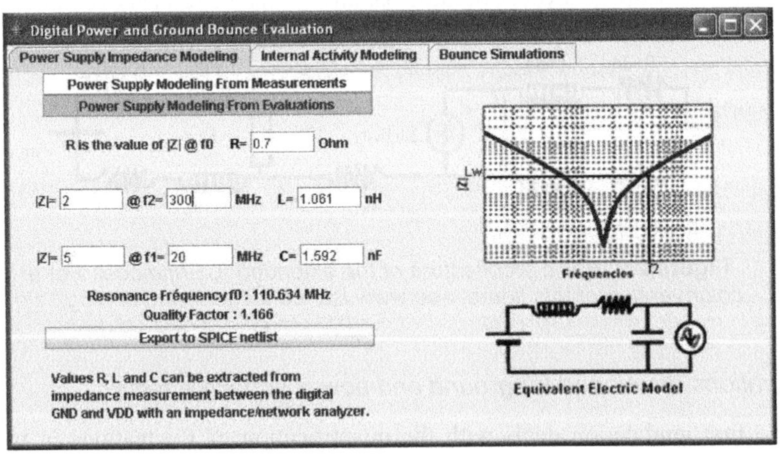

b)

Figure 2.10. *Equivalent digital capacity; this depends on the technology (number of gates per mm²) and the numerical area. Typical values are 0.15–0.3 nF/mm² for standard 0.35 µm CMOS technology and 0.8–1.6 nF/mm² for 90 nm. These passive parasitic elements can be estimated or determined from the measurement/parameter. For a color version of this figure, see www.iste.co.uk/gontrand/analog2.zip*

Second, the internal activity of the keypad is described in a simple way. This current source is the superposition of periodic current pulses. The characteristics of each pulse (height, width, frequency, rise and fall times) can be easily determined from the general parameters of the digital circuit: its average consumption, its clock frequencies, its upward and downward transition times, its inclinations and each clock latency (Figure 2.11).

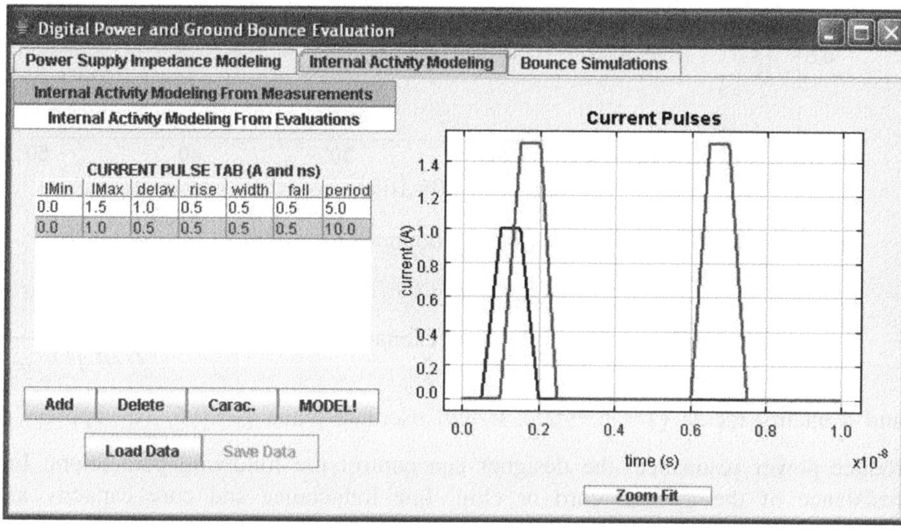

Figure 2.11. *Dedicated application to quantify digital bounces of ground and power (internal activity). For a color version of this figure, see www.iste.co.uk/gontrand/analog2.zip*

The last step concerns the ground and power bounce calculations. This is done in the time domain (analytical resolution of the waveforms per part) or in the frequency domain (Fourier analysis). Figure 2.12 shows an example of simulated V_{DD} power supply resonance for different line resistors (from 0.05 to 0.5 Ω).

Future trends should take into account in-depth contacts via Green methods to invest, for example, spic (Smart Power ICs). Basic rules for reducing the resonance of the digital power supply network with the tools are presented as follows: the user can estimate how the analog ground will fluctuate due to the activity of the digital block(s), although power supplies are completely separate (Figure 2.13). Moreover, these tools can help to understand the malfunctions of digital blocks with an unstable digital power supply.

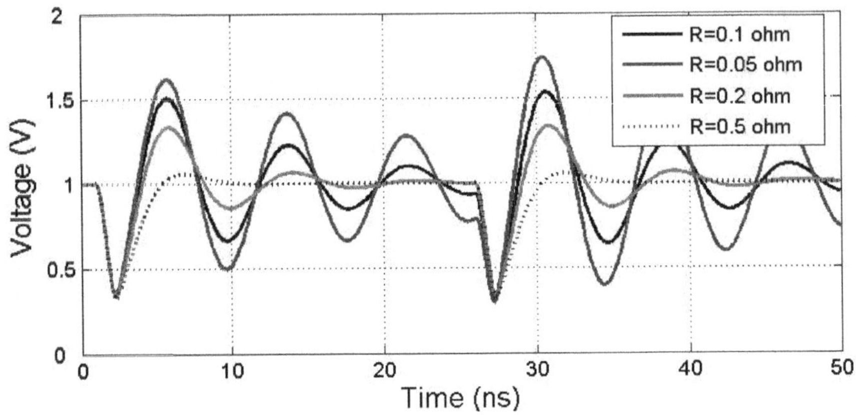

Figure 2.12. *Simulated power supply bounces for different line resistances*

The passive distribution network has a resonant frequency $f_{res} = {}^{1}\!/_{2\pi\sqrt{LC_{core}}}$ and a quality factor $Q = {}^{L\omega_{res}}\!/_{R}$. It will oscillate when Q > 0.5. To suppress or reduce power resonance, the designer can control the following parameters: line resistance of the internal card or chip, line inductance and core capacity and additional decoupling capacity on the chip. Below, we summarize the influence of these parameters on the resonance of the digital power supply:

– It leads to a decrease in the oscillation amplitude without any influence on the oscillation frequency (Figure 2.12).

– The increasing oscillation L leads to a decrease in the oscillation frequency without any influence on the amplitude, and increasing C_{core} leads to a simultaneous decrease in the oscillation amplitude and the oscillation frequency.

An intuitive way to decrease power and bounce to ground is to add a huge decoupling capability to the PC3 power grid. In fact, this external decoupling capacitor only filters the parasitic voltage of the power grid and is not enough to significantly reduce interference. The main disturbing frequency is due to the resonance of the "RLC" path composed of the inductance and resistance of the bonding wire and the capacitance of the core of the digital part of the circuit. Increasing off-chip capacity as much as possible will not avoid on-chip natural resonance. Another simple solution is to add a series resistor to the power supply network (on the PCB or inside the chip).

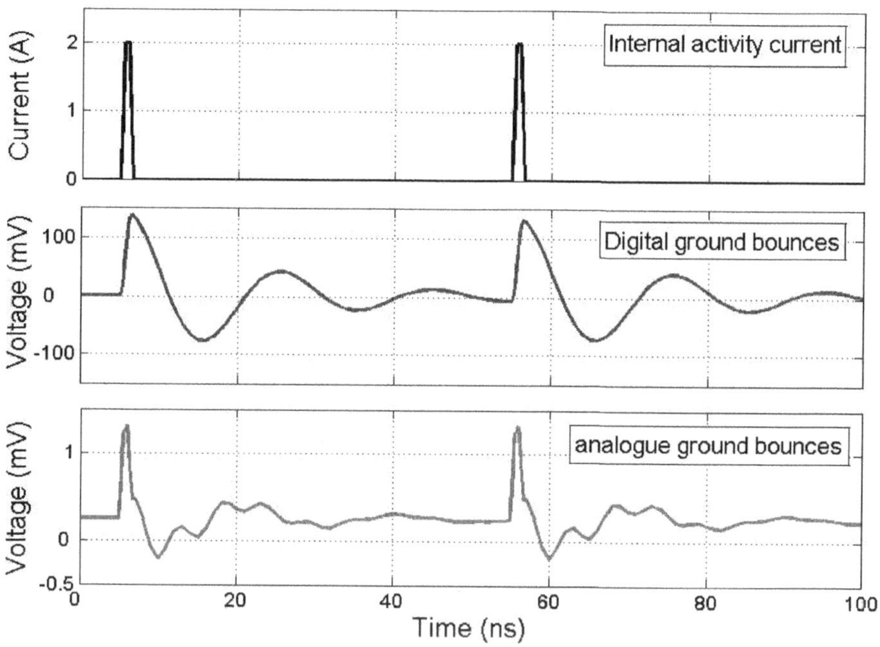

Figure 2.13. *Simulated analog mass fluctuations*

The added resistance decreases the quality factor of the RLC network and, at the same time, dampens the induced oscillations. The low-pass filter obtained gives good results. But too much electrical network resistance greatly increases the chip's power consumption and may cause functional defects. These are the reasons why this type of noise reduction technique is not recommended for a large digital circuit but can be applied to a small digital application such as a frequency synthesizer.

Internal activity (equivalent digital current source) also has a direct link with the power and amplitude of ground oscillation. For clock frequency below resonance frequency, power supply resonance will be improved when resonance frequency is harmonic with clock frequency (worst case is for $f_{res} = 2f_{clock}$). When resonant frequency is below clock frequency, the passive distribution network acts as a low-pass filter. An increase in the average digital current consumed leads to an increase in the oscillation amplitude. A decrease in transition times leads to a decrease in the oscillation amplitude.

Insulation of the substrate from the noisy digital block to a sensitive analog block is discussed in the next section.

Comparative substrate insulation study on a virtual test case study

In this section, we perform a comparative study of different substrate insulation solutions applied to a virtual integrated circuit test case. The characteristics of the virtual circuit are as follows: 130 nm technology in a highly resistive substrate (10 Ω·cm) is used; the digital and analog areas are both 10 mm^2 and the "packaging" used is a thin quad flat panel display with 100 surface wires of 14 × 14 mm^2. Some typical values can be evaluated from these parameters: digital core capacitance is 7 nF and each bonding wire induces a parasitic resistance and an inductance of 120 mΩ and 6 nH, respectively. A first-order electrical model can be developed from these simple assumptions and can be useful for technology and implementation choices.

The following is a list of solutions that can be used to reduce the impact of substrate noise in such a device:

– choice of substrate (high or low resistivity), rear bulk connection (grounded or floating), number of $V_{DD\text{-}GND}$ power supply pairs, number of links, arrangement protection ring around the block, and triple well arrangement (Figure 2.14). Good efficiency of guard rings or triple wells is achieved by polarizing each structure with a specific feed/mass platform. The triple well technique would offer the best substrate insulation at low frequencies. The N-well must be as small as possible in order to decrease the area of the reverse biased diode, and therefore the capacitance between the analog or digital part and the substrate. A small capacity can filter higher frequencies.

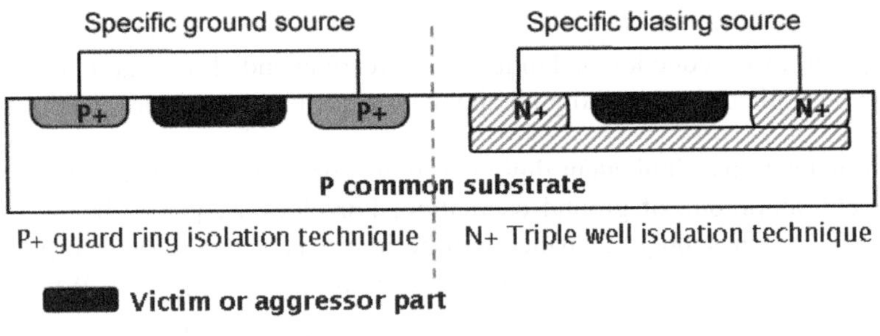

Figure 2.14. *Disposition substrate insulation techniques: guard ring and triple well*

Simulated insulation is quantified by the transfer function between the differential analog power supply and the digital internal activity in d3. Here, the different configurations used are as follows:

– first case is the reference: no layout protection, high resistance substrate with floating matrix bottom;

– second case: the back of the substrate is grounded;

– third case: the high resistance substrate is replaced by a low resistance substrate (bottom of the floating substrate);

– fourth case: the high-resistance substrate is replaced by a low-resistance substrate (grounded back substrate);

– fifth case: the noisy digital block and the sensitive analog block are surrounded by two guard rings;

– sixth case: the sensitive analog block is insulated in a triple well;

– seventh case: the noisy digital block is insulated in a triple well.

Figure 2.15 shows different transfer functions. A first resonance around 50 MHz, corresponding to the resonance frequency of the digital power supply network, is observed on these different waveforms. The second resonance peak is about 200 MHz and corresponds to the resonance frequency of the analog power supply network.

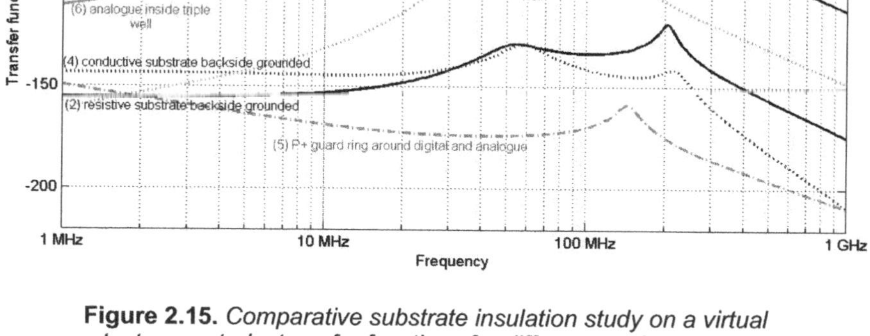

Figure 2.15. *Comparative substrate insulation study on a virtual test case study: transfer functions for different configurations*

The best insulation technique to reduce substrate coupling for this virtual integrated circuit is to surround the digital block and the analog block with two protection rings. The worst solution is to use a low-resistance substrate with the floating back. This result cannot be generalized to other designs as each case is specific. The electromagnetic coupling between connecting wires is also not taken into account in this simple virtual study. This explains the excellent insulation provided by the P+ guard ring technique: guard rings are polarized with perfect Kelvin points in this case and dissipate perturbations of the substrate as well as possible. In addition, triple-well insulation can be very effective in insulating a small-area block, but now considering a large-area block, the parasitic capacitance between the N-well and the substrate will be larger and the cutting frequency smaller.

2.3. Noise in integrated oscillator circuits

2.3.1. *Oscillator design considerations*

It is well known that modern telecommunication systems require full integration in the substrate using a standard process. Millimeter wave generation is a key microwave function for transceivers. Analyzing and designing high-frequency oscillators requires more creativity than logic. At high frequencies, parasitic capacitances, wire inductances and other parasitic components also change their reactance and oscillation frequency, feedback amplitude, output power and other AC variables.

However, there is no generic method for designing a low-phase noise oscillator. Usually, scientists use their in-house “know-how” and the low-phase noise character is achieved only by using a prototype that is difficult to reproduce for industrial implementation. In addition, oscillator characteristics such as power consumption, output power and adjustment range are very important and are also optimized. It is therefore very important to develop a generic method that will allow us to achieve a good precision of design of integrated circuits.

Hereinafter, we present an analysis and design methodology of a 20 GHz low-phase noise VCO (voltage-controlled oscillator) using a 0.25 µm BiCMOS SiGe:C technology, developed by ST Microelectronics. Possible applications include broadband fiber optic reception circuits with 20 Gb/s and 40 Gb/s, or broadband satellite and radar transmissions in the Ka-band (the name used during World War II, for reasons of discretion): 20/30 GHz.

2.3.2. *VCO topography*

Two topographies based on SiGe:C-HBT devices were compared for the designs presented: VCO-LC, using a cross-coupled differential pair, and a balanced common base, Colpitts VCO. Figure 2.16 shows the differential structure using crossed transistors.

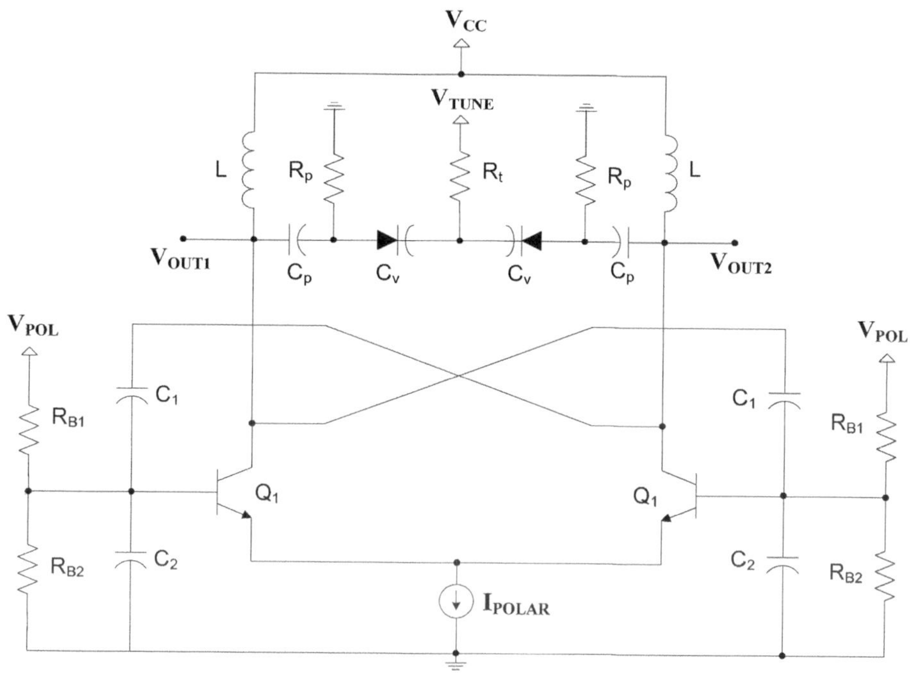

Figure 2.16. *Topography of the VCO using a cross-coupled differential pair*

It has positive feedback caused by a capacitive divider (C_1, C_2) crossed between the base and collector terminals of the transistors of the differential pair. A major concern in the design of a bipolar junction differential pair oscillator is to avoid forward biased base collector junctions (see saturation), which would significantly increase tank (sink/tank) losses and phase noise. In fact, only a fraction n – see equations [2.1] and [2.2] – of the tank collector voltage is fed back to the respective cross-coupled base through capacitive dividers C_1–C_2 (which can be used to adjust the oscillation amplitude but modifying, at the same time, the phase noise generated by the bipolar junction transistors).

These capacitive connections allow for independent biasing of the collector of Q_2 (Q_1, respectively) of the base of Q_1 (Q_2 respectively), thus allowing control of the voltage swing on the base of the transistors. Switching transistors associated with the positive feedback provide a negative resistance that compensates for resonator LC losses and thus leads to an oscillating regime. Cp capacitances are used to reduce varactor nonlinearities. The first step of the oscillator simulation analysis is to study the stabilization of the circuit, according to the Barkhausen criterion; the ADS software offers the "OscTest" tool to study this condition through the parameter S11.

Oscillation frequency depends mainly on the values of the "LC" components in the resonator as well as on the capacitive bridge C_1, C_2 and parasitic capacitances in transistors, such as:

$$f_{osc} = \frac{1}{2\pi\sqrt{LC_{eq}}}$$, where f_{osc} is the oscillation frequency. [2.1]

and, with a very good approximation:

$$C_{eq} = \frac{C_1 C_2}{C_1 + C_2} \qquad [2.1b]$$

Operating conditions of transistors are determined by the feedback capacitance ratio n and the bias current I_{POLAR}, with

$$n = \frac{C_1}{C_1 + C_2} \qquad [2.2]$$

Three modes of operation (dynamic modes) can be distinguished: linear, nonlinear and highly nonlinear. The "slightly nonlinear" mode allows optimization of the output voltage swing and a relatively low harmonic level. At a fixed I_{POLAR}, the capacitance ratio n controls operating conditions of transistors of the differential pair. A low capacitance relationship increases the gain compression of the transistors, and then supports the influence of their nonlinearity, transconductance and capacitances (junction capacitance, transition capacitance and cross-capacitances). This mode of operation also reduces the conduction time of transistors. Usually, this last point is a key to reducing phase noise, due to the cyclo-stationary properties of the noise sources associated with the collector current. Figure 2.17 shows the phase noise at an offset of 1 MHz versus the capacitance ratio n.

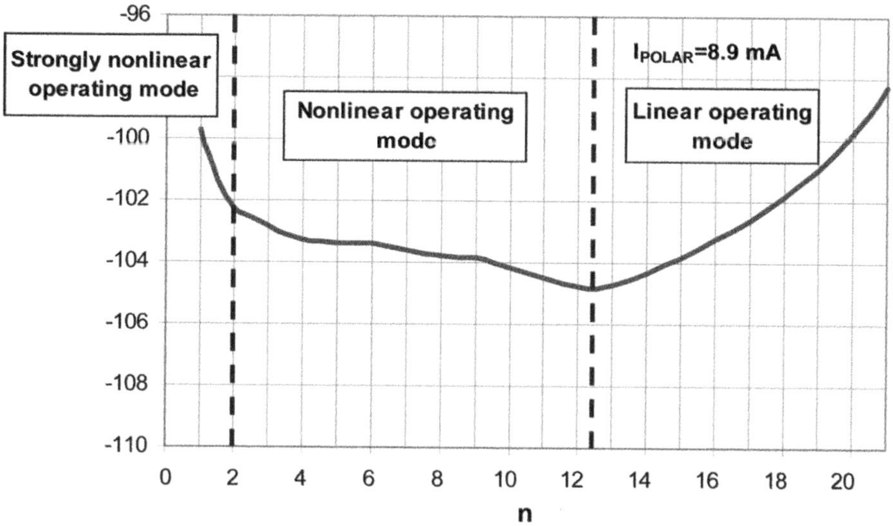

Figure 2.17. *Phase noise at 1 MHz (dBc/Hz) offset (carrier frequency is close to 20 GHz) versus capacitance ratio n (bias current: I_{POLAR} = 8.9 mA). The THBT used is NN232A128. For a color version of this figure, see www.iste.co.uk/gontrand/analog2.zip*

A very nonlinear region can be observed for $2 < n < 12.5$. When n is greater than 2, we see an improvement in phase noise, moving away from the nonlinearities of transistors that deteriorate this phase noise. However, an increase in noise is noted if n is too low ($n < 2$). The phase noise then tends to increase while the conduction time of the transistors decreases.

In the next section, we discuss the main optimization performances: spectral output power, oscillation frequency, phase noise, output impedance, bandwidth, second harmonic rejection and power consumption. Figure 2.18 presents the VCO diagram of a varactor-adjusted LC Colpitts differential topography in a common base configuration.

This circuit can oscillate at frequencies higher than those of the common emitter Colpitts oscillator because its high frequency gain is higher. The fundamental frequency signal is collected at differential outputs Vout1 and Vout2. The second harmonic can be probed directly at the common node of R_{B1} and R_{B2}. Oscillators with a cross-coupled differential pair use an LC resonator circuit that imposes a limit on the maximum possible oscillation frequency.

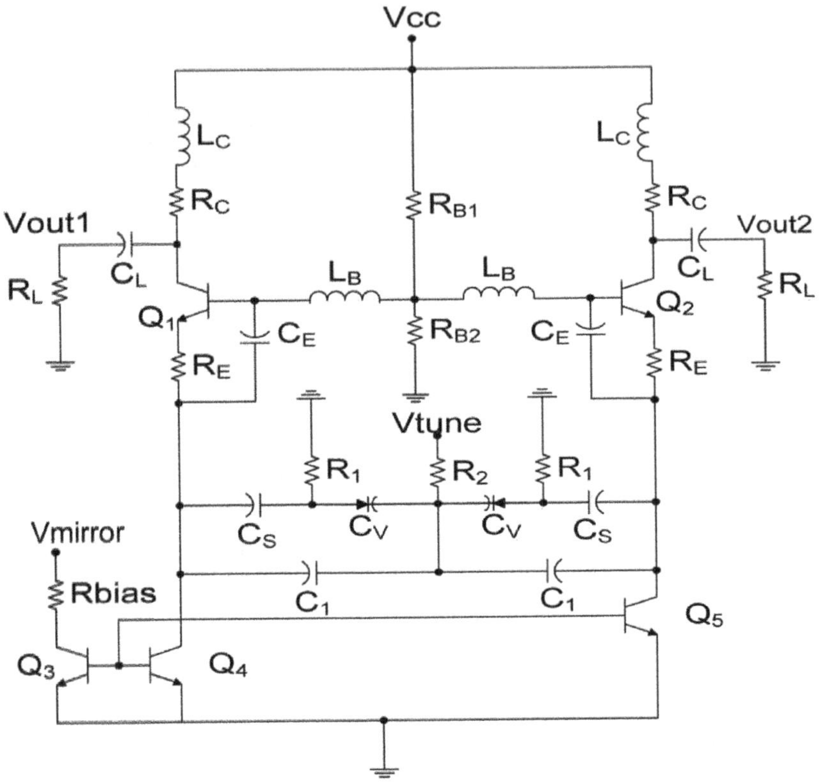

Figure 2.18. *Topography of a Colpitts oscillator*

In addition, the resonance tank is charged directly by the buffer, thereby increasing the current in the resonator. The reduction in size is an advantage of the common design of basic Colpitts compared to crossed differential pair oscillators (nearly half an area per part). In addition, the tank circuit, located in the base of the transistor, is easily insulated from the collector charge. This architecture ensures low noise by minimizing the number of transistors in the circuit. The negative resistance is the result of a capacitive charge (see varicap diode C_V) of the emitter. Designing as large an inductor as possible, thus maximizing the Q-factor, is also beneficial for high tank impedance (with the smallest bias current). The transistor handles negative resistance buffers and at the same time the tank from the charge. With this low-noise topography, the minimization of phase noise is a key point; in this way, capacitors C_E, C_V and inductance L_B are maximized. In addition, the transistor is biased to the minimum noise current density for the oscillation frequency, and the size of the emitter is adjusted to obtain the best noise match to the tank impedance. R_E degeneration resistance is included to suppress harmonics and reduce up-converted noise (although some thermal noise is added).

It is a low-phase noise and low-power consumption design.

An important issue in the design of high-performance oscillators is to obtain a high-quality LC resonator. The quality factor has a significant impact on phase noise performance. Also, we can use the equations as follows:

$$\frac{1}{Q}=\frac{1}{Q_{R_p}}+\frac{1}{Q_{R_L}}+\frac{1}{Q_{R_C}} \qquad [2.3]$$

$$L\{\Delta\omega\}=10\log\left\{\frac{kT\cdot[1+F]\cdot\left(\frac{\omega_0}{\Delta\omega}\right)^2}{\frac{A^2}{2}\cdot\omega_0 CQ}\right\} \qquad [2.4]$$

where $L\{\Delta\omega\}$ is the spectral density of the phase noise at the shift frequency $\{\Delta\omega\}$ of carrier ω_0, F represents the excess noise of the differential pair of transistors, A is the amplitude of the differential output voltage and Q is the total quality factor of the resonator. Q_{RP}, Q_{RL} and Q_{RC} represent the quality factors related to parasitic inductance resistance, inductance value and circuit capacitance, respectively. These components must be carefully designed in order to obtain the desired oscillation frequency f_0. To determine the value of Q, it is necessary to take into account the size of the inductor and thus seek a compromise between a high value Q and a smaller dimension of the inductor.

Although, for low frequencies, Q is mainly given by inductance series resistors, for high frequencies this is not the case, as the inductance size is smaller. This corresponds to a very small winding and therefore a very small resistance in series. The real resonant capacitor is composed of the varactor as well as the active charge elements, and the inductor's parasitic or fixed capacitances. In practice, the most effective way to reduce phase noise is to use a resonant tank with a high-quality factor. In addition, it is advisable that the phase noise level does not depend on the resonant frequency and is therefore constant with the control voltage.

The energy consumption of the VCO should at least compensate for the losses of the resonant tank. Therefore, it is not surprising that power losses decrease linearly for smaller series resistors in the resonator. We note that to satisfy both the phase noise and in the optimized power consumption design, it is necessary to obtain an LC resonator with maximized ratios L/R and L/C. Using these conditions and our design tools, we have achieved the optimized design of a 20 GHz VCO that we show in detail in the next section.

2.3.3. *Results and discussion*

The two different VCOs operate with supply voltages of 2.5 V. The topography of the crossed differential pair provides a constant bias current of 4.45 mA (in the case of balanced Colpitts, it is equal to 2.50 mA). Low secondary harmonic rejection and high output power instability were observed. Incorporating a buffer improves the VCO output characteristics: good insulation between the oscillator and the charge (50 W) to reduce the pull factor in the circuit, linear operation of the transistor on the buffer to control the distortion of the output signal and an input impedance that varies very little with frequency, which guarantees constant level output power over the VCO setting range and better power consumption.

To optimize VCO functionality, two configurations (follower emitter and common emitter) were studied and compared. The second harmonic rejection was not sufficient (about 19 dB) to meet the initial requirements (25 dB) and therefore a simple bandpass filter was placed in the buffer output. This filter increases the rejection of the second harmonic. At the same time, it retains the initial performance of the VCO. To avoid degradation of small fundamental output power variations, a wide filter adjustment range and a second highly efficient harmonic filtering are required. The common emitter configuration has a higher output power (7 dBm) than the following transmitters (4 dBm).

The output power is almost constant on the setting voltage for both buffer configurations. A good setting range value was obtained, that is, 395 MHz and 420 MHz for the follower emitter and the common emitter, respectively. However, the common configuration of the emitter uses an additional inductance of 1.26 nH, which leads to a larger size of the final circuit. The push and pull factors were also simulated in SPectreRF and ADS by varying the supply voltage and charge resistance, respectively, and observing the oscillation frequency change. The arrangement of the LC-VCO cross-coupled differential pair occupies 430 μm × 480 μm chip area and consumes 22.3 mW at 2.5 V. Particular attention has been paid to make the arrangement of the differential circuits as symmetrical as possible. Figure 2.19 shows a typical VCO.

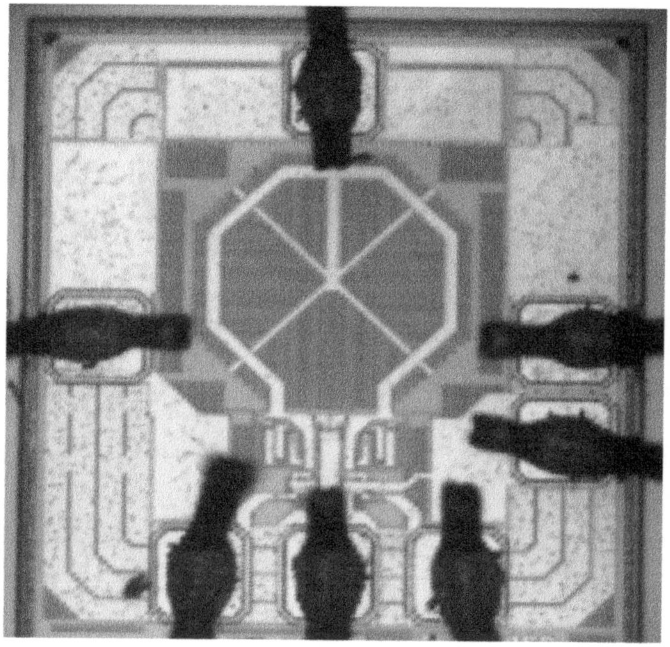

Figure 2.19. *Photograph of a VCO (core: 0.25 μm). For a color version of this figure, see www.iste.co.uk/gontrand/analog2.zip*

The spectrum of single-ended output power at a center frequency of about 20.483 GHz (SiGeC process, core: 0.25 μm) is shown in Figure 2.20.

In addition, the arrangement of the inductors and their placement relative to the core of the VCO have been carefully evaluated. The distance between the inductors has also been optimized, with respect to magnetic coupling. Finally, the two VCOs display very good performance in terms of phase noise (–106 and –96.5 dBc/Hz at 1 MHz offset), power consumption (with 22.3 mW at 2.5 V power supply without output buffer contribution), output signal power (4 dBm and 1.1 dBm, respectively, by the two topographies) and second harmonic rejection of 27 and 25 dBm (with an output filter for cross-coupled differential pair topography).

Depending on the phase noise performance, very good levels were noted for both topographies. In Figure 2.21, the measurements of phase noise as a function of frequency are plotted, in the case of the topography of the cross-coupled differential pair.

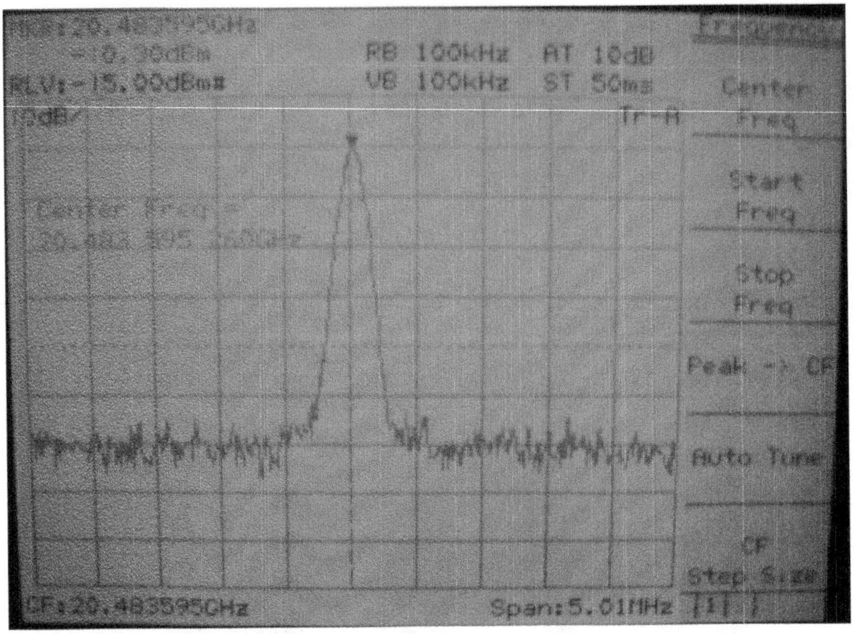

Figure 2.20. *20 GHz VCO spectrum. For a color version of this figure, see www.iste.co.uk/gontrand/analog2.zip*

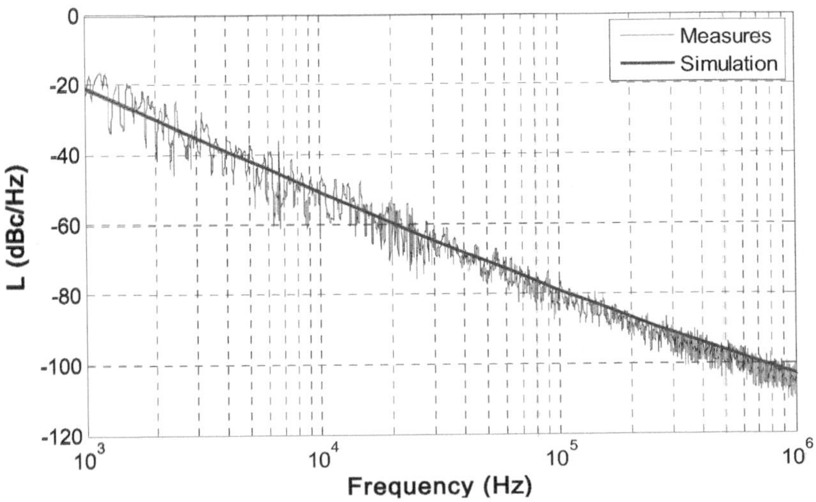

Figure 2.21. *VCO phase noise. For a color version of this figure, see www.iste.co.uk/gontrand/analog2.zip*

In addition, the oscillation frequency with respect to the setting voltage is shown in Figure 2.22.

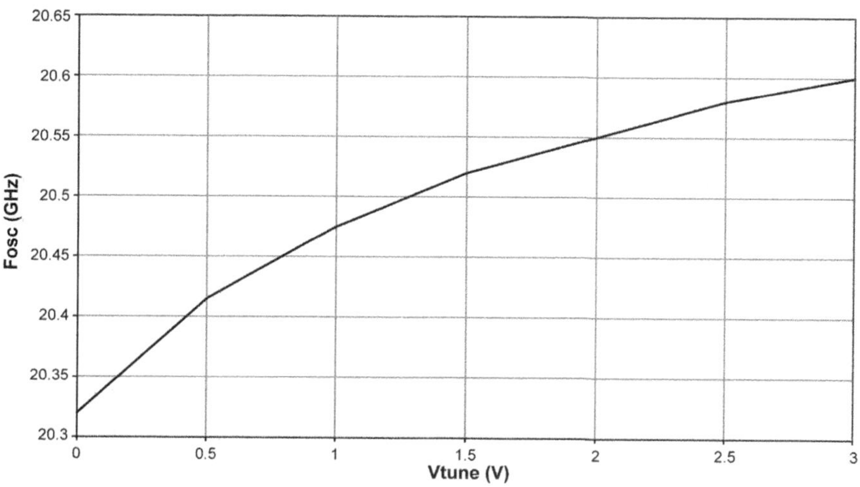

Figure 2.22. *Oscillation frequency versus setting voltage*

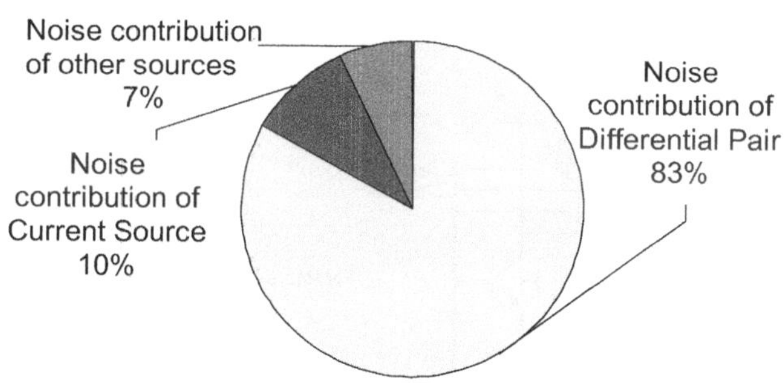

Figure 2.23. *Noise distribution. For a color version of this figure, see www.iste.co.uk/gontrand/analog2.zip*

The adjustment voltage can vary from 0 to 3 V to obtain a carrier frequency in the range 20.32 – 20.60 GHz (Δf = 280 MHz). Throughout this range, the simulations and phase noise measurements are very similar. We had already developed this methodology in a BiCMOS SiGe technology of 0.35 µm with a good agreement between the simulated and measured data for a VCO of 5 GHz. A low-frequency noise model using an original method based on the concept of correlation resistance has been developed and associated with a non-quasi-static nonlinear transistor model. For both VCOs, the dominant LF noise is due to the noise of the differential pair (~83%, Figure 2.23).

It is mainly generated by the transfer current noise and internal base resistance noise for the 20 GHz VCO (~75%) of transistors Q_1 and Q_2. There is a reduction in the noise contribution of the differential pair when we increase the number of base fingers, resulting in a reduction in the base resistance. Nevertheless, increasing the base fingers also produces an increase in energy consumption; for this reason, the number of fingers should not be large. To compare VCOs in different frequency bands and with different energy consumption, we define a figure of merit given by the following equation:

$$\text{FOM} = L_{meas}\left(f_{offset}\right) - 20 * \log\left(f_{osc} / f_{offset}\right) + 10 * \log\left(P_{consumption} / 1\text{mW}\right) \quad [2.5]$$

In this equation, $L_{meas}(f_{offset})$ is the phase noise measured at a frequency offset f_{offset} from the center frequency f_{osc}, and $P_{consumption}$ is the VCO power consumption in milliwatts. Using this definition, a FOM of −179 and −172 is obtained for the two oscillators studied.

Sensitivity functions of a VCO

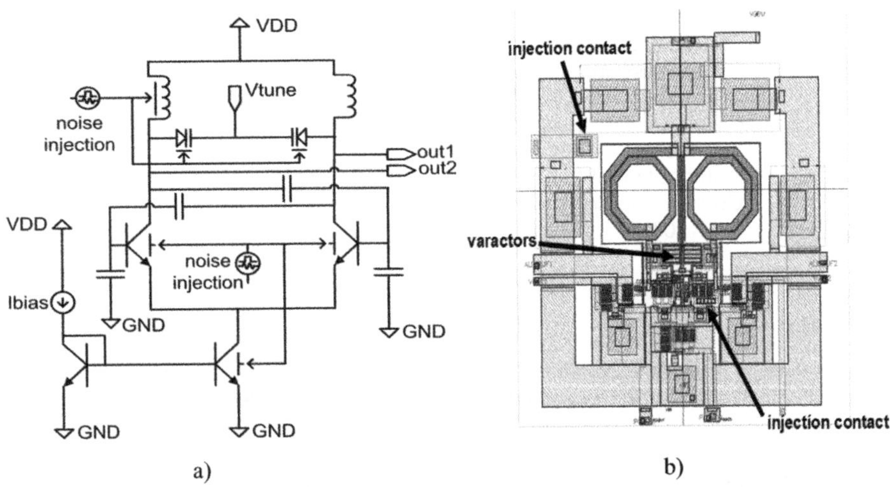

Figure 2.24. *(a) Core of the VCO with contact injection; (b) VCO layout*

Now, we will engage in a quantitative description of VCO behavior: we focus on the sensitivity of VCOs to parasites injected into the substrate. This part is essentially experimental (but is followed by a theoretical part). The following VCO, operating near 5 GHz, was designed in a high resistivity BiCMOS (SiGe) process technology of 0.35 (µm). The main part of the VCO is the LC tank. A coupled and crossed differential pair of bipolar heterojunction transistors biased by a current I_{bias} induces negative resistance.

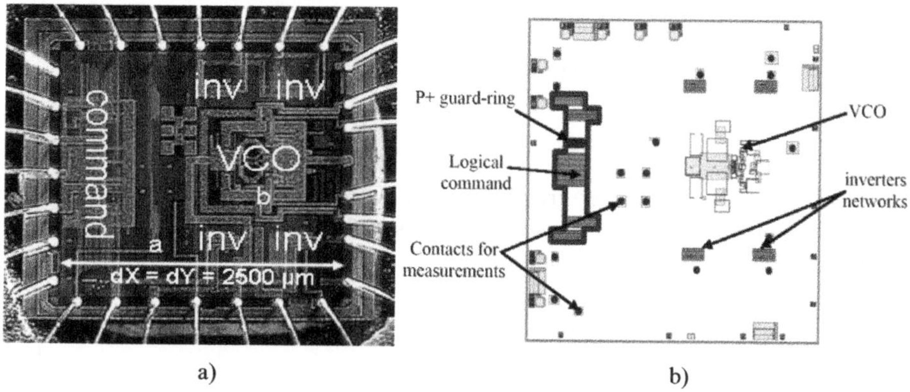

Figure 2.25. *(a) Microphotography of an evaluation circuit; (b) a layout of the test chip with substrate contacts and buried layers*

In order to inject a parasitic signal directly into the substrate with an external generator, we placed several contacts near the differential transistors and inductors (Figure 2.25). The supply voltage of the circuit is 3.3 V. The adjustment voltage can vary from 0 to 4 V to obtain a carrier frequency of 4.25–4.60 GHz. The output signals pass through buffers formed by a common collector circuit configuration. The power delivered to a 50 Ω charge under this current is close to 0 dBm. The phase noise measured at 100 kHz from the carrier varies from 90 dBc/Hz up to 100 dBc/Hz (depending on the dispersion and polarization of the chip).

To avoid frequency variations due to supply voltage noise, many decoupling capacitors are placed in the VCO. For the same purpose, substrate contacts and protection rings around the functional blocks that have a large ground contact area have been added. Other substrate contacts with different areas and locations are placed on the test chip to allow us to estimate the resistance of the substrate.

2.4. Sensitivity functions

The VCO carrier depends on the bias current or the control voltage. The variations of VCO carrier function of the bias current or of the control voltage can be described by certain sensitivity functions called K_{bias} and K_{tune}, respectively (see below). A change in the I_{bias} current can modulate the carrier frequency in the same way that a change in the V_{tune} voltage can simultaneously change the varactor capacitance and the output frequency. In this context, consider a sine wave disturbance superimposed on the bias current or control voltage:

$$v_m(t) = A_m \cos(\omega_m t) \tag{2.6}$$

The carrier frequency changes according to the expression:

$$f = f_c + K \cdot v_m(t) \tag{2.7}$$

where f_c is the VCO carrier frequency and K is the sensitivity function. After frequency integration to determine the phase, the oscillator output signal was determined by the following equation:

$$s(t) = A\cos\left(\omega_c t + \frac{A_m K}{f_m}\sin(\omega_m t)\right) \tag{2.8}$$

The modulation index ($A_m K/f_m$) depends on the sensitivity function. For a supply voltage perturbation, the modulation index has a low value, so we can assume in our calculations that the modulation band is reduced to a narrow range. We refer to the carrier in order to calculate the perturbation level in the spectral domain. With these assumptions, the spectral power with respect to the carrier in dBc units is given by:

$$P_{sbc} = 20 \cdot \log\left[\frac{K \cdot A_m}{2 f_m}\right] \tag{2.9}$$

It measures the difference between the parasitic sideband power and that of the carrier. A perturbation superimposed on the bias current or the control voltage generates sidebands whose power amplitude in dB is directly proportional to the sensitivity K (equation [2.6]). At the same time, the sensitivity K is a function of the bias current and the control voltage. Thus, the noise power under the carrier varies when the bias or control is changed. In order to compare the sensitivity functions and their effects on the power below the carrier when a substrate disturbance is applied, we consider the evolution of the frequency with respect to the bias current and the control voltage. The sensitivity function is obtained by deriving the frequency curve (see Figures 2.26 and 2.27).

Sensitivity function simulations were carried out with the Cadence Spectre RF simulator (Cadence n.d.). The perturbation effect on the bias current causes a small variation in the carrier frequency. The voltage applied to the varactors generates a much higher level of sensitivity than the sensitivity to the bias current. Basically, the amplitude of the sensitivity function for the bias current K_{bias} is 10 times smaller than the setting voltage sensitivity K_{tune}.

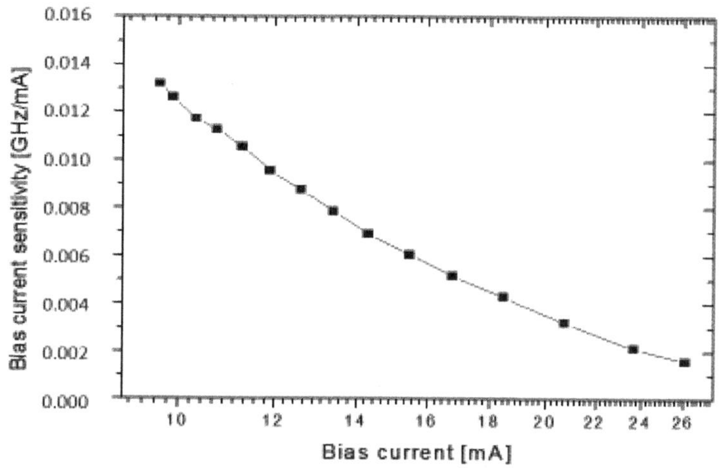

Figure 2.26. *Bias current sensitivity function (K_{bias})*

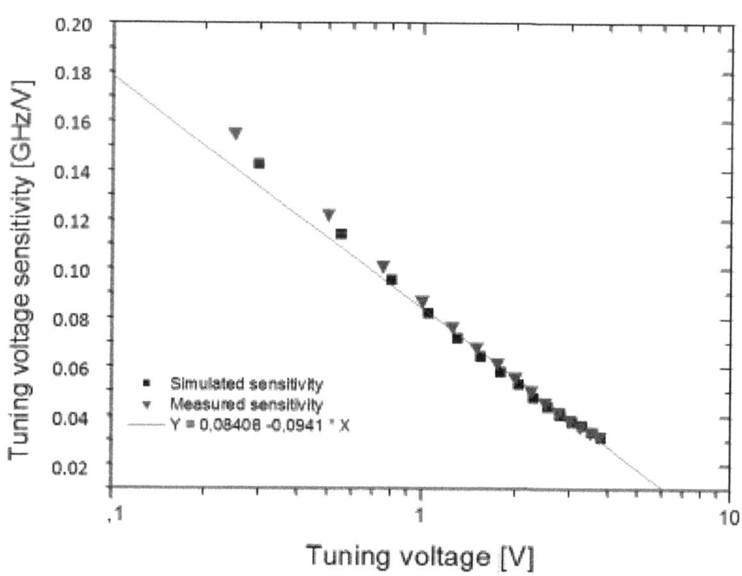

Figure 2.27. *Adjustment voltage sensitivity function (K_{tune}). For a color version of this figure, see www.iste.co.uk/gontrand/analog2.zip*

This method allows us to establish the sensitivities of the VCO depending on the bias circuit conditions using measurements and/or simulations. The impact of a perturbation on the bias source or the control voltage can easily be evaluated from equation [2.8]. These assumptions could reduce the impact of power supply noise. VCO polarization must be correctly chosen in order to decrease the sensitivity to the VCO frequency.

Parasitic sidebands of the VCO due to substrate perturbations

Experimentally, we measure the closed VCO parasitic sidebands at the carrier frequency f_c when the sine wave voltage signal is injected into the substrate (Figure 2.28).

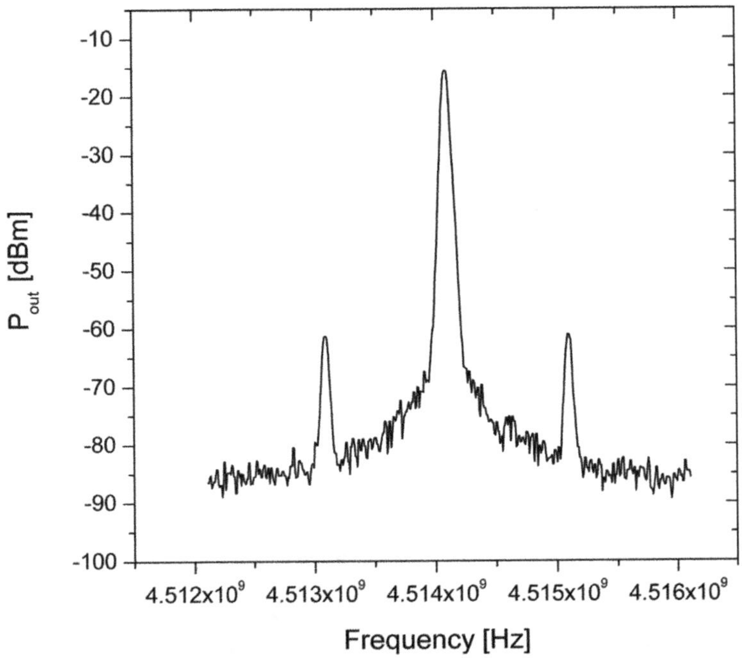

Figure 2.28. *VCO spectrum (process node: 0.25 mm) sensitivity due to the injected perturbation (level, location, frequency, etc.) is quantified by Psbc values (Figures 2.29 and 2.30)*

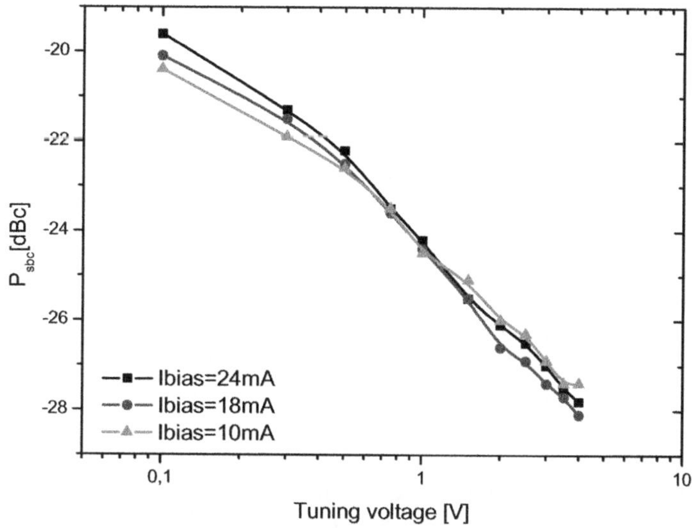

Figure 2.29. *Measured Psbc function of the control voltage (50 m Vpk at 100 kHz perturbation is injected near the inductors). For a color version of this figure, see www.iste.co.uk/gontrand/analog2.zip*

Substrate noise therefore induces power supply bounces V_{ss} and V_{DD}. In our specific test chip, taking into account the high resistivity substrate, noise attenuation with distance, the substrate cannot be considered as a single node, hence, many on-chip decoupling capacitors and injection nodes inside the VCO. We obviously have a direct path between the injected nodes and the volume of the passive or active device (varactor, inductance, heterojunction bipolar transistor [HBT], etc.). This is the point that we analyze by studying the sensitivity of components indicating the contribution of each to the output noise. Simulated transfer functions between the source directly applied to the pins of the component substrate and the VCO output were studied.

Simulated data for a Psbc show that devices that have a large N-well and/or N-underground layer area (varactors and transistors) make a significant contribution to output noise. A correlation between sensitive components and Psbc values has been established. Injecting the noise of the sine wave near the substrate of the bipolar junction transistor generates large parasitic sidebands. An equivalent impact is achieved when the same sine wave is injected into both the bipolar junction transistor and the varactor substrate. The impact of the bipolar substrate noise exceeds the contribution of the varactor for similar injection conditions. Some of the elements of the substrate based on measurements between the injection contacts and ground were determined to accurately simulate the impact of a sinusoidal disturbance of the substrate. Passive and active device models, which are used for our simulations, contain equations for parasitic coupling elements. In this way, to compare the measured and simulated data, it is necessary to implement elements describing this.

The contacts used to inject noise into the substrate have different surfaces and are placed in different locations. Circuit components have substrate insulations around them (protection ring resistance) of different resistance values, connected to ground. The resistance of the contact of the injection substrate with the ground depends on its location due to parasitic elements. Thus, the injected perturbation will be more attenuated when the substrate insulation (protection ring resistance) of a device has a low resistance value. With these observations and our measurement interpretations, we created a basic substrate model by adjusting resistances of a grounded guard ring and grounded substrate, respectively, for injection into the transistor, inductance and varactor substrate. Figure 2.30 shows the power of the sideband under the Psbc carrier as a function of the control voltage sensitivity. Psbc measurements and simulations were obtained using a peak injection voltage of 150 mV. This sine wave was injected near one of the inductors.

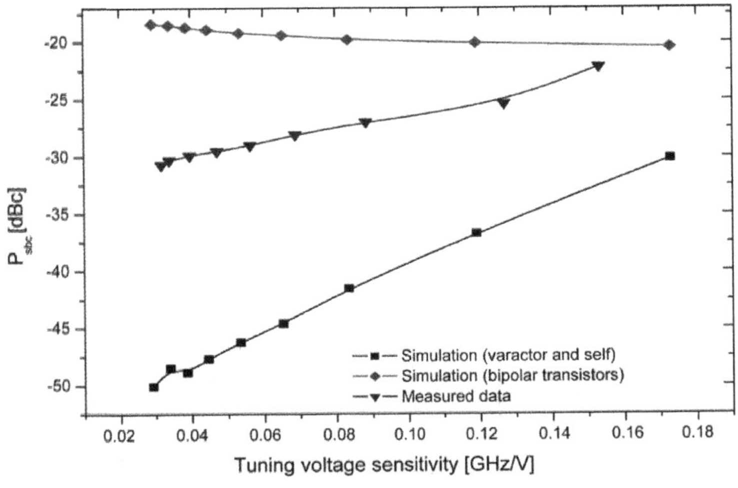

Figure 2.30. *Control voltage sensitivity. For a color version of this figure, see www.iste.co.uk/gontrand/analog2.zip*

Circuit simulations with our basic substrate model show in particular that two main devices can convert substrate noise into output power: transistors and varactors. Noise converted by transistors with respect to the control voltage sensitivity is a roughly constant function and contributes significantly to the Psbc level. On the other hand, noise converted by varactors leads to a significant decrease in Psbc with the control voltage sensitivity. Comparing simulated data obtained with a simple substrate model and measured data shows that Psbc values are dominated by the behavior of the varactor (Psbc increases according to the level of sensitivity of the control voltage). With injection near one of the inductors and considering the attenuation of the substrate, the sidebands are mainly introduced by the sensitivity of the varactor to the harmonic noise of the low-frequency substrate.

With the injection close to transistors of the cross-coupled pair, measured Psbc values are higher than those of Psbc for close inductance injection (for the same injected perturbation) and do not depend on control voltage sensitivity. This observation is consistent with simulated behavior obtained with our simple substrate model.

2.5. New developments in impulse sensitivity function

2.5.1. *Oscillators: brief recap of the theory*

In this section, we will discuss in detail the theoretical analysis of sensitivity functions. We "build" an LC Colpitts oscillator using a mixed-mode simulation (see below). Autonomous circuits are not easy to study, especially with regard to signal algorithms (small or large). Since an oscillator operates from a DC power supply and is not driven, the phase of any oscillation on the simulation is not uniquely defined. Without loss of generality, a schematic circuit of an LC-tuned oscillator may be as shown in Figure 2.31.

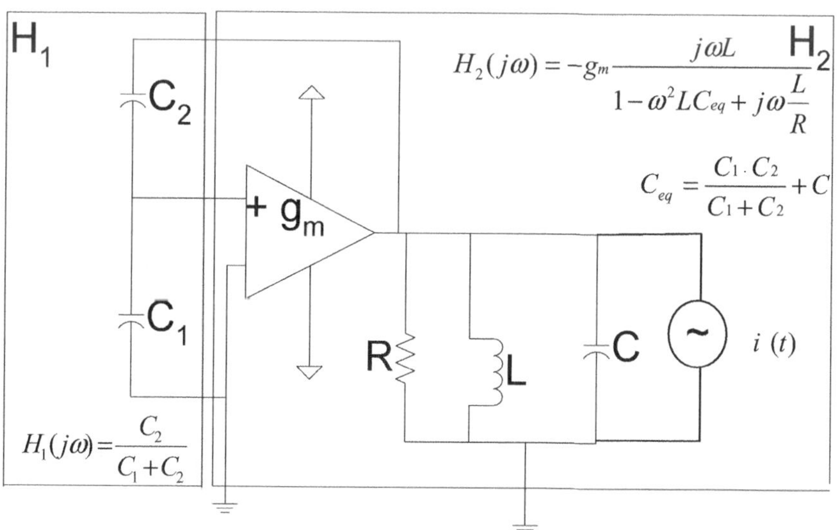

Figure 2.31. *Colpitts oscillator diagram*

A small signal analysis of circuits (see Barkhausen criterion: mod(H1*H2) = 1, gives an oscillation frequency:

$$\omega_0 = \frac{1}{\sqrt{L \cdot Ceq}}, \text{ where } C_{eq} = \frac{C1 \cdot C2}{C1 + C2} + C \qquad [2.10]$$

The admittance of a parallel "RLC", for $\Delta\omega << \omega_0$, can be written as:

$$Y(\omega_0 + \Delta\omega) = G_L .(1 + \mathrm{j}.2.Q_L .\frac{\Delta\omega}{\omega_0}) \qquad [2.11]$$

where G_L is the parallel parasitic conductance and Q_L is the effective quality factor of the entire tank. For a steady-state oscillation, transconductance $g_m(t)$ is equal to 1/R. Then, it is simple to show that the output impedance transfer function (closed loop) of the oscillator is defined as follows:

$$\alpha_0 = \frac{v_{out}(\omega_0 + \Delta\omega)}{i_{in}(\omega_0 - \Delta\omega)}$$

It is the inverse of the imaginary part of Y, that is:

$$\alpha_0 = (2.\mathrm{j}.G_L.Q_L .\frac{\Delta\omega}{\omega_0})^{-1} \qquad [2.12]$$

Intrinsic perturbation of Colpitts sources can be defined as, for example, a voltage source in series with the inductor or current sources in parallel with the capacitor (see continuity of current and voltage, respectively, when applying a parasitic pulse) – a voltage source in series either with the supply voltage or with the base of the transistor, for example.

After running dozens of oscillation periods, this relevant equilibrium state is stored and then followed by several transient simulations that run after a perturbation pulse is applied. In the case of the diagram presented above, the noise transfer functions (NTFs), at certain nodes of interest, can then be derived as follows:

$$\begin{aligned}
\alpha_0^{(C_2,i)} &= \frac{1}{2}\times\frac{C_1}{C_2+C_1}\times\frac{1}{2j\omega_0 C_{eq}}\left(\frac{\omega_0}{\Delta\omega}\right)\\
\alpha_0^{(+,0)} &= \frac{1}{2}\times\frac{C_2}{C+C_2}\times\frac{1}{2j\omega_0 C_{eq}}\left(\frac{\omega_0}{\Delta\omega}\right)\\
\alpha_0^{(i)} &= \frac{1}{2}\times\frac{1}{2j\omega_0 C_{eq}}\left(\frac{\omega_0}{\Delta\omega}\right)\\
\alpha_0^{(C_1/C_2,0)} &= \frac{1}{2}\times\frac{C_1+C_2}{C_1}\times\frac{1}{2jQ}\left(\frac{\omega_0}{\Delta\omega}\right)
\end{aligned} \qquad [2.13]$$

(the left factor 1/2, to the right of equations [2.13], comes from the oscillation condition, which leads to v+ = v-, and then, by calculating v+ + v-, implies these expressions for a one-sided transfer function). $N_\phi = N_s.(\alpha_0)^2$ where N_s is the spectral density (of the considered source), and K is a division coefficient dependent on C_1 and C_2. For example, if the noise source can be represented by a pulse ("dirac") of amplitude I0, its spectral density will be constant with respect to the frequency (see white noise), and we obtain:

$$\begin{aligned}
N_{s_{2,i}} &= \left(\frac{1}{C_{eq}}.\frac{C_1}{C_2+C_1}\right)^2\times\left(\frac{I_0}{4\Delta\omega}\right)^2\\
N_{s_{+,0}} &= \left(\frac{1}{C_{eq}}.\frac{C_2}{C_1+C_2}\right)^2\times\left(\frac{I_0}{4\Delta\omega}\right)^2\\
N_{si(t)} &= \left(\frac{1}{C_{eq}}\right)^2\left(\frac{I_0}{4\Delta\omega}\right)^2\\
N_{s_{1/2,+}} &= \left(\frac{C_1+C_2}{C_1}\right)^2\times\left(\frac{\omega_0}{Q}\right)^2\left(\frac{I_0}{4\Delta\omega}\right)^2
\end{aligned} \qquad [2.14]$$

As shown in Figure 2.31(a), the behavior of the HBT device is modeled with microscopic equations (see drift diffusion model (DDM)) and it is "surrounded" by passive elements (L, C, R), governed by Kirchhoff's laws; this is the famous mixed simulation mode.

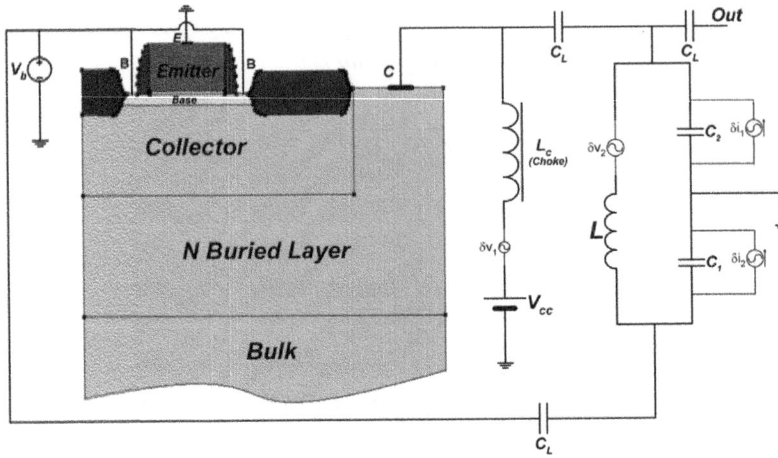

a)

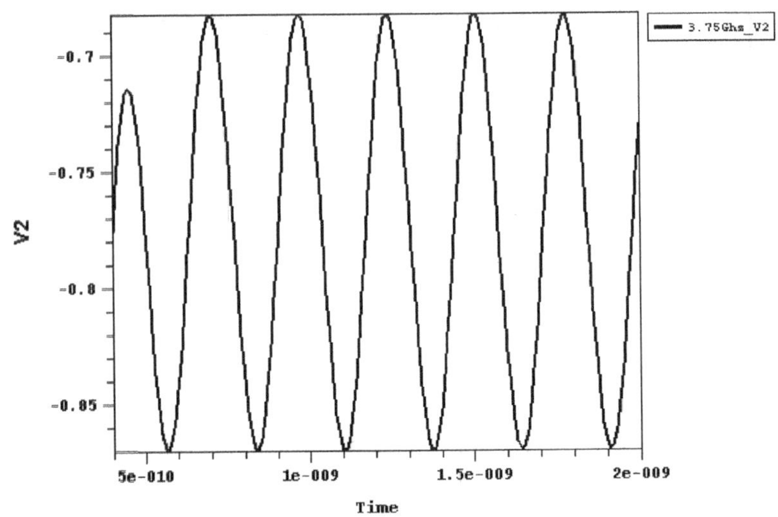

b)

Figure 2.32. *(a) Device diagram (DDM) surrounded by passive elements (SPICE). (b) A typical HBT (mixed-mode simulation) oscillator response: voltage at the collector node. For a color version of this figure, see www.iste.co.uk/gontrand/analog2.zip*

Figure 2.32(b) presents a typical case of oscillation: T = 268 ps or f = 3.75 GHz; a pulse was not needed to start the oscillation. The transistor gain and the operating point have been chosen in a simple way to obtain a very stable oscillation.

One of our main goals is to explore and test the impulse sensitivity function (ISF) using our microscopic approach, potentially expanding its applicability. Hajimiri states that capturing phase noise involves parasitic pulses affecting the tank, and the sine wave without a DC value crosses the zero value.

2.5.2. *Pulse sensitivity function (with some recap)*

In this section, we discuss the linear time variant model for analyzing VCO sensitivity to substrate perturbations. As previously mentioned, the method is based on the calculation of the ISF, which represents the excess phase after the application of a parasitic pulse on an oscillator circuit. If we inject a current pulse as shown in Figure 2.32(a), the instantaneous voltage change Δq V is given by:

$$\Delta V = \frac{\Delta q}{C_{tot}}$$

where Δq is the total injected charge due to the current pulse (its integration over time) and C_{tot} is the total capacitance between nodes of interest (amplifier output and ground). If we "ignore" the resistor and represent the parasitic pulse current noise modeled as a "dirac", we get an equation of the following type:

$$\frac{1}{\omega_0^2}\frac{d^2V}{dt^2} + V = \frac{\Delta q}{C}\delta(t) \qquad [2.15]$$

Note that the current pulse only changes the voltage across the capacitor.

The ISF function $\Gamma(\tau)$ can be accessed by varying the delay of the Dirac perturbation pulse event during an oscillator period. In the first stage of the book, we did not consider amplitude change because it can disappear over time; on the other hand, the phase shift is preserved. The perturbation pulse may be a current spike in a capacitive node (injected charge) or a voltage spike in an inductive node.

$\Gamma(\tau)$ is dimensionless and has the same period as that of the oscillator (or we can consider that $\Gamma(\omega_c\tau)$ has a period of 2p). Finally, the excess phase due to a Dirac pulse applied at time t can be written as:

$$h_\phi(t,\tau) = \frac{\Gamma(\omega_0\tau)}{q_{\max}}u(t-\tau) \qquad [2.16]$$

where $u(t)$ is the unit step and q_{max} is the maximum charge through the capacitor C_{tot}: ($q_{max}=C_{tot}*Vout_{max}$).

In other words, $\Gamma^{\phi}(\tau)$ is a direct representation of the excess phase (phase shift), normalized by injected charge (for a current pulse). Due to its periodicity, $\Gamma^{\phi}(\tau)$ can be developed in series as follows:

$$\Gamma^{\phi}(\omega_c\tau) = \frac{c_0}{2} + \sum_{n=1}^{\infty} c_n \cdot \cos(n\omega_0\tau + \theta_n) \qquad [2.17]$$

Let us now consider the harmonic parasitic signal (current or voltage) defined by its amplitude and its angular frequency ω_p (frequency: f_p), that is:

$$p(t) = A \cdot \cos(\omega_p t) \cdot u(t - t_0) \qquad [2.18]$$

The phase shift is directly obtained according to the following expression:

$$\phi(t) = \int_{-\infty}^{t} \Gamma^{\phi}(\omega_0\tau) \cdot p(\tau) \cdot d\tau = \int_{t_0}^{t} \Gamma^{\phi}(\omega_0\tau) \cdot A \cdot \cos(\omega_p\tau) \cdot d\tau \qquad [2.19]$$

The phase shift can be developed as follows:

$$\phi(t) = A \cdot \frac{c_0 \sin(\omega_p t)}{2\omega_p} + A \cdot \sum_{1}^{\infty} \frac{c_n \sin\left[(n\omega_0 \pm \omega_p)t + \theta_n\right]}{2(n\omega_0 \pm \omega_p)} + \phi_0(t_0) \qquad [2.20]$$

If $\omega_p << \omega_0$, then $\phi(t)$ is approximated by the first term (in other words, the terms in $n\omega_0$, for all n, disappear).

$$\varphi(t) = Ac_0 \frac{\sin(\omega_p t)}{\omega_p} \qquad [2.21]$$

Equation [2.28] indicates that higher order terms of the ISF are important for perturbation signals around oscillator harmonics: for closed perturbations at harmonic n, only the term is calculated with $\omega_p = n\omega_0 + \Delta\omega$ (where $\Delta\omega << \omega_0$) and must be taken into account in the final expression of the oscillator output $s(t) = S \cdot \cos(\omega_0 t + \varphi(t))$, while at n = 0, all other terms are negligible. Finally, only low-frequency noises ( ) and perturbations around harmonics ( ) influence the phase. The coefficient of the ISF function is used to calculate the phase shift introduced by the pulsation noise $\omega_p = n\omega_0 + \Delta\omega$. We recall that the coefficient c_n depends on the location and is taken into account in the final expression of the oscillator output, while all other terms are negligible. Finally, only low-frequency noise (n = 0) and noisy perturbations around harmonics (n = 1,2,...) influence the phase. Coefficient c_n of the ISF function is used to calculate the phase shift introduced by the noise at the pulse $\omega_p = n\ \omega_0 + \Delta\omega$. Then, the oscillator output can be written for small values of φ (using basic trigonometric formulas, plus a first-order serial expansion):

$$V_s(t) = V_0 \cos[\omega_0 t + \varphi] \approx V_0 \cos[\omega_0 t] - \varphi \sin[\omega_0 t] \quad [2.22]$$

By identification, we have:

$$V_s(t) = V_0 \cdot \left[\cos(\omega_0 t) - \frac{c_n A}{2\Delta\omega} \sin(\Delta\omega t) \cdot \sin(\omega_0 t)\right] \quad [2.23]$$

According to a narrow sideband condition, noise results in a pair of sidebands at $\omega_0 + \Delta\omega$, with a sideband power with respect to the carrier, equal to:

$$P_{sbc}(\Delta\omega) = 10 \cdot \log\left[\left(\frac{c_n A}{4\Delta\omega}\right)^2\right] \quad [2.24]$$

Note that we obtain a unique P_{sbc} (in dB_c) expression [2.24] for any harmonics, by using the $c_0/2$ term in [2.20], instead of c_0. This result can be compared to equation [2.24].

Note that we get a unique expression (in dBc) for P_{sbc} v c_0.

2.5.2.1. *Impulse sensitivity: fresh insights*

Now, we would like to extend the discussion to any phase value (ϕ∈ [0, 2π]), not just for a small charge injection.

If we consider a parasitic pulse injection (a "dirac") at time τ, the overall output voltage of the oscillator can be considered as the superposition of the DC voltage and the response to the parasitic injection, because any solution shifted in time remains a solution:

$$V_T(t) = V_s(t) + p(t) = V_0 \cdot \cos(\omega_0 t) + \frac{q}{C} \cdot \cos(\omega_0 t) \cdot u(t-\tau) \quad [2.25]$$

This equation can be rewritten as:

$$V_T(t) = V_{T0} \cos[\omega_0 \tau + \varphi] \quad [2.26]$$

where:

$$V_{T0} = \sqrt{V_0^2 + (\frac{q}{C})^2 + \frac{2qV_0 \cos(\omega_0 \tau)}{C}} \quad [2.27]$$

and:

$$\varphi = \arctan(\frac{-q.\sin(\omega_0 \tau)}{CV_0 + q.\cos(\omega_0 \tau)}) \pm k\pi \quad [2.27b]$$

We obtain the ISF by deriving [2.27b]:

$$\dot{\varphi}(\tau) = \frac{-q.(CV_0 \cos(\omega_0 \tau) + q)}{C^2 V^2{}_0 + q.(q + 2CV_0.\cos(\omega_0 \tau))} \quad [2.28]$$

Now, utilizing these overarching formulas and assuming a limited expansion of ϕ, we get:

$$\varphi = \frac{-q.\sin(\omega_0 \tau)}{CV_0}(1 - \frac{q.\sin^2(\omega_0 \tau)}{3.(CV_0)^2}) \quad [2.29]$$

If we consider the first order, we get:

$$\varphi = \frac{-q.\sin(\omega_0 \tau)}{CV_0} \quad [2.30]$$

If we compare the latter formula to equation [2.29], we get the well-known result:

$$\Gamma = -\sin(\omega_0 t), \text{ with } q_{max}=CV_0 \qquad [2.31]$$

This result is valid, but for very small q. Note that a small signal analysis cannot give any information on the amplitude of the oscillation, as it is determined by the nonlinearities of the system. In a real circuit, the oscillation frequency depends on the junction capacitances of the transistor (which can no longer be managed by our mixed-mode simulations). In Figure 2.33, we represent q extracted from simulations in mixed mode and compared to the "arctan" analytical solution of equation [2.34b], where C is given by $C = \dfrac{C_1 \cdot C_2}{C_1 + C_2}$.

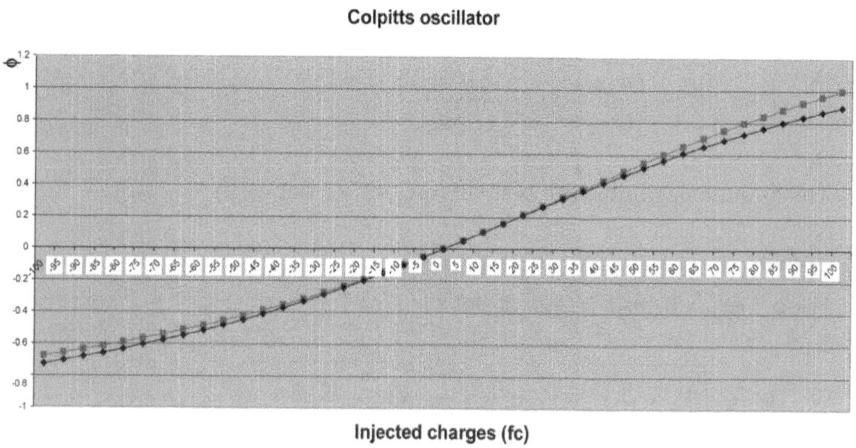

Figure 2.33. *Phase shift (at collector) versus injected charges (mixed mode compared to our analytical model: equation [2.27b]. For a color version of this figure, see www.iste.co.uk/gontrand/analog2.zip*

In this typical case, "mixed mode" CV0 = 100 fc is in very good compliance with the "arctan" adjustment of equation [2.27b], considering the graphical asymmetry (offset) along the y-axis. For this characteristic calculation, we chose an almost –90° angle of "high amplitude" (meaning not exactly 90°) to "observe" the influence of the denominator of equation [2.27b].

To our knowledge, this is the first time such comparisons have been presented. There is twofold interest: our "microscopic" simulations seem to be validated by our analytical formulations, and we do not need to stay in the linear domain, as in ISF theory where only a very limited domain of Φ versus q is considered. In fact, to have consistent results, using SPICE simulations, the actual charge injected due to the actual sources of noise and interference is supposed to be several orders of magnitude smaller than the quantities of charges injected, which is too restrictive, into the actual circuits. According to equations [2.27] and [2.27b], the phase and/or amplitude may change after the perturbation. In Figure 2.34, we represent mixed-mode simulations when applying a pulse on V_{DD} (the shock inductor has been replaced by a resistor). Our observations are as follows.

When crossing zero

For $\phi = +/- \phi_{max} = \arctan(+/- q/CV_0)$, $\omega_0\tau == +/- \pi/2$, or $\tau = T/4$ or $3T/4$.

When crossing zero in the downward direction ($\tau = T/4$): $\phi = -\phi_{max}$. When crossing zero in the upward direction ($\tau = 3T/4$): $\phi = \phi_{max}$.

$$\Delta V = \sqrt{V_0^2 + (\frac{q}{C})^2} - V_0 \sim \frac{q^2}{2C^2V_0} \qquad [2.32]$$

This is quite negligible if q is small ($q/C << V_0$, our case).

At peaks

For $\phi = \phi_{min} = 0$, $\omega_0\tau = \pi$ or π, but $\tau = T/$or T:

$$\Delta V = \sqrt{(V_0 + \frac{q}{C})^2} - V_0 \text{ or } \Delta V = \sqrt{(V_0 - \frac{q}{C})^2} - V_0 \qquad [2.33]$$

This means that we have:

$$\omega_0\tau = 2\pi\text{:} \ \ \Delta V = +\frac{q}{C} \qquad [2.34]$$

and:

$$\omega_0\tau = \pi\text{:} \ \ \Delta V = -\frac{q}{C} \qquad [2.34b]$$

typically with $\Delta V = 1\text{–}5$ mV, implying that $|q| = 5\text{–}15$ fc, which is good in the linear domain of f versus q.

Our "raw" simulations (q_0 – pulse, negative) show these results (see Figure 2.34), with good numerical accuracy; here, we consider the relationship $|\Delta V| = \frac{|q|}{C}$.

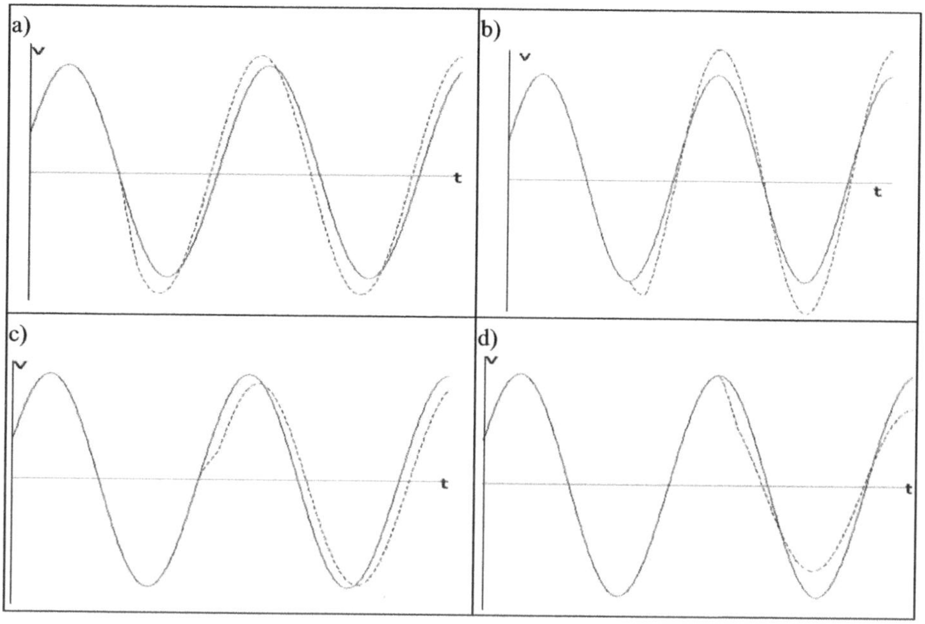

Figure 2.34. *Mixed-mode simulations: (a) crossing zero when descending, (b) bottom, (c) zero in the upward direction and (d) peak*

Figure 2.35 shows other results concerning the ISF at certain nodes of interest (base, collector, C_2: see Figure 2.36).

Sine waveforms are observed in each numerical experiment, as predicted by the theory; it would be wise to do an exhaustive study on such simulations. We started this book, adapting (with some success) equation [2.27] with numerical simulations. Finally, we have dealt with some cases where the phase shift can be very significant. We achieved this, for example, by applying a high parasitic voltage pulse to a node of the tank capacitor, injecting roughly the same charge as the total charge present on the capacitor plate. In the delay time range τ of [0, T], the variation of f is linear by parts.

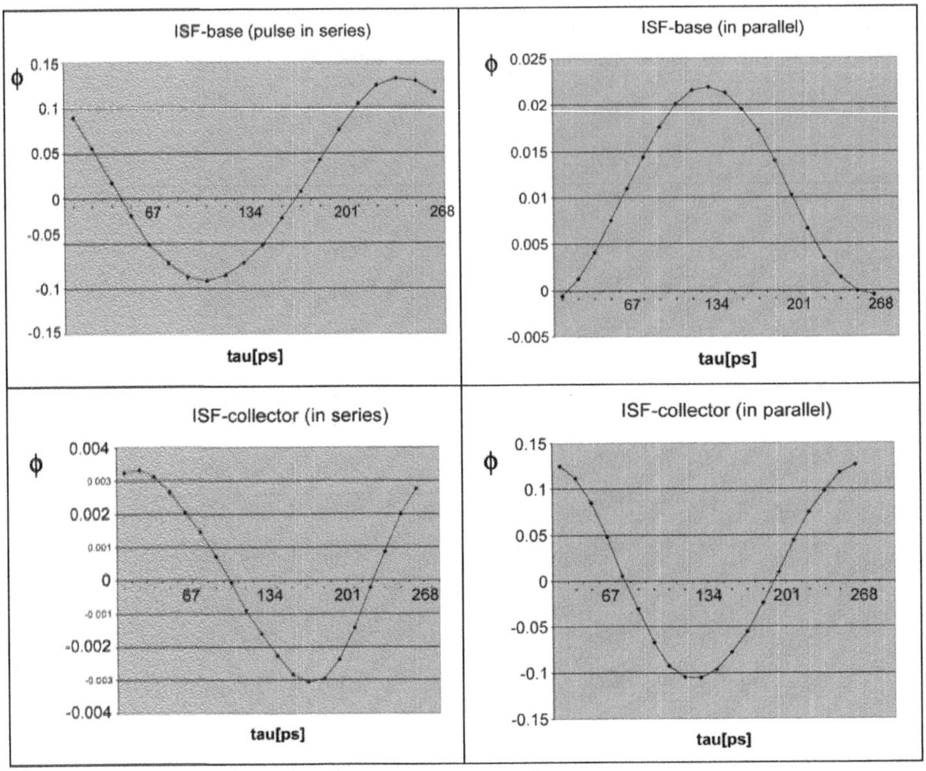

Figure 2.35. *Phase shifts at different nodes of interest*

As shown in Figure 2.36, in the linear range, $\omega_0/2$ (with an accuracy of a few percent: 2 or 3), which is in very good agreement with equation [2.29] (when $q = CV_0$). To go further, an interesting question is how to calculate phase noise, when Dirac pulses are applied.

Fluctuations $\frac{\varphi}{\omega_0}(\tau)$ are the cause of phase noise, depending on the noise sources (e.g. process-induced traps for HBT: ambient noise).

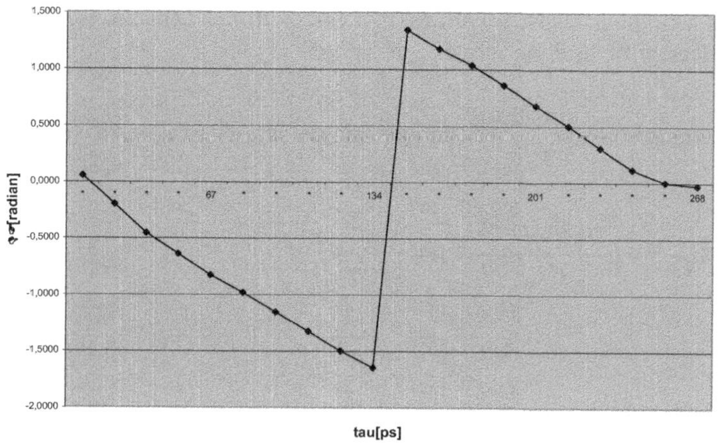

Figure 2.36. *Phase shift for high pulse at tank capacity node*

If we are interested in temporal fluctuations, we can study the following expression:

$$\tau_{\varphi} = \frac{\varphi(t+\tau) - \varphi(t)}{\omega_0} \quad [2.35]$$

Taking the square of its RMS value:

$$\sigma^2 =< {\tau^2}_{\varphi} >= \frac{< \varphi(t+\tau) >^2 - 2 < \varphi(t)\varphi(t+\tau) > + < \varphi(t) >^2}{\omega_0^2} \quad [2.36]$$

If we now say that this process is a stationary second order process, the RMS value does not depend on time:

$$\varphi_{RMS}^2 =< \varphi(t)^2 >=< \varphi(t+\tau)^2 >$$

The second term to the right of the equation (equation [2.36]) is the well-known correlation function: $C_{\varphi}(\tau)$.

Equation [2.36] is reduced to:

$$\sigma^2 =< {\tau^2}_{\varphi} >= 2\frac{(\varphi_{RMS}^2 - C_{\varphi}(\tau))}{\omega_0^2} \quad [2.37]$$

Let us take the parasitic waveform that comes from a capacitive charge injection, at the frequency f0, as done above.

We have: $\varphi = \dfrac{-q.\sin(\omega_0\tau)}{CV_0}$ (for low values of ϕ; otherwise, take equation [2.36]):

$$\varphi_{RMS}^2 = <(\frac{-q.\sin(\omega_0\tau)}{CV_0})^2> = \frac{q^2}{2C^2{V_0}^2} \quad [2.38]$$

and:

$$C_\varphi(\tau) = \varphi_{RMS}^2.\cos(\omega_0\tau) \quad [2.38b]$$

or:

$$\sigma^2 = \frac{q^2}{C^2{V_0}^2\omega_0^2}(1-\cos(\omega_0\tau)) \quad [2.39]$$

which gives the associated noise power density:

$$S_\varphi(\varpi) = \frac{q^2}{2.C^2{V_0}^2\omega_0^2}\left(2.\delta(0) - [\delta(\omega-\omega_0) + \delta(\omega+\omega_0)]\right) \quad [2.40]$$

The last objective is to calculate the power spectral density (PSD) of the oscillator via the autocorrelation output voltage:

$$S_V(\omega) = FT(\frac{1}{t}\lim_{t\to\infty}\int_{-\frac{t}{2}}^{\frac{t}{2}} <V(t+\tau).V^*(t)> dt) = FT(V_0^2$$
$$\lim_{t\to\infty}\int_{-\frac{t}{2}}^{\frac{t}{2}} .<e^{j\omega_0\tau}e^{.(\varphi(t+\tau)-\varphi(t)}> dt) \quad [2.41]$$

It is easy to verify that the variance satisfies relationships:

$$\frac{d^2}{d\tau^2}[\sigma^2(\tau)] = C_\varphi(\tau);\ \frac{d}{d\tau}[\sigma^2(0)] = 0;\ \sigma^2(0) = 0 \quad [2.42]$$

With these conditions and given that φ(t) varies slowly over time, equation [2.41] can be approximated as follows:

$$S_V(\omega) = FT(V_0^2 e^{jw_0\tau} e^{\frac{-\omega_0^2.\sigma^2(\tau)}{2}}) \text{ (FT: Fourier transform)} \quad [2.43]$$

Here, σ^2 is given by [2.39].

Thus, equation [2.43] can be rewritten as follows:

$$S_V(\omega) = FT(V_0^2.\exp^{-\frac{q^2}{2C^2V_0^2}}.\exp^{j\omega_0\tau}\exp^{\frac{q^2.\cos(\omega_0\tau)}{2.C^2V_0^2}}) \quad [2.44]$$

Finding the Fourier transform in equation [2.44] is equivalent to finding the Fourier transform of the following:

$$\exp^{\frac{q^2.\cos(\omega_0\tau)}{2.C^2V_0^2}}, \text{ i.e.: } S'_V(\omega) = 2\int_0^\infty \cos(\omega\,\tau)\exp^{\frac{q^2.\cos(\omega_0\tau)}{2.C^2V_0^2}}d\tau \quad [2.45]$$

This general formula (numerical solution), can be reduced, for the specific case where q<<CV0 (equation [2.44]) can be re-written as:

$$S'_V(\omega) = \frac{2}{T}\sum_{n=1}^{\infty}\int_0^T \cos(n\omega_0\tau).(1+\frac{q^2.\cos(\omega_0\tau)}{2.C^2V_0^2})]d\tau \quad [2.46]$$

It can be reduced to:

$$S'_V(\omega) = \delta(0) + \frac{q^2}{4C^2V_0^2}.[\delta(-\omega_0) + \delta(\omega_0)] \quad [2.47]$$

Finally, we have considered a few cases where the phase shift can be very significant. We achieved this when applying a source of parasitic high-voltage pulses in series with the tank inductor or capacitor. In the delay time range t of [0,T], the variation of f is linear with respect to τ, but seems to have a discontinuity at both ends of this range (Φ seems to tend toward –π and + π [tank inductance: ISF]; we obtained the same curve, concerning capacitor C_2 – not shown here).

The slope, in the linear range, for both cases – inductance and capacitor – is the same: $\omega_0/2$ (with an accuracy of a few percent), in the range [$-\pi/2, +\pi/2$]. This identity does not seem like a coincidence, but it is not so easy to explain. In fact, if ϕ versus τ was linear from $-\pi$ to π, the slope should be twice ω_0 and $\phi = \omega_0\,\tau$ and should be extracted from equation [2.27b] for high injections, that is, CV0 << max (q.cos ($\omega_0\tau$)) (with discontinuities for ω_0 that give a zero denominator, or when $\omega_0\tau = +/- \pi/2$). At this point, it is important to present a possibly original view of such oscillators: the concept of virtual damping; it is considered a "phase diffusion", as a physical diffusion of particles, called Brownian motion. The phase noise of the oscillator has been studied from an in-depth mathematical point of view.

In this case, the response of the Γ- function is not appropriate (the system is neither linear nor quasi-linear), but is broadened to a finite bandwidth. Oscillator phase fluctuations are quite analogous to the Brownian motion of a free particle; therefore, the phase fluctuation can become quite large, without any energy cost; the spectral line width that involves the phase fluctuations cannot be processed by quasi-linear methods.

The amplitude of the signal from the output of the damped oscillator takes the form (Hajimiri 1999):

$$V_s(t) = V_0 \exp(-Dt) \cdot \cos(\omega_0 t + \varphi) \qquad [2.48]$$

(assuming Gaussian white noise: $<\phi^2> = 2Dt$);

where D can be called "broadcast frequency".

By taking N (with N very large) identical oscillators, with the same initial condition (e.g. $\phi = 0$ at $t = 0$), each oscillator loses its initial phase information after a "sufficient time".

We postulate that any oscillator will be damped, and stopped, in a week, or a century, if there is no intrinsic noise that opposes this phenomenon. We recall that Langevin found the formula called the Einstein Brownian diffusion coefficient: "from Newton's law, and by introducing a friction force for the particle, it will stop... in a given time: but, it is a noise (which he assumed to be white, with a zero average value), which calls into question the phenomenon of slowing down". In "real life", intrinsic white noises with zero mean value should exist in or near the circuits; their correlation function can be written as:

$$< \eta(\tau_1).\ \eta(\tau_2) > = A^2\ \delta_{ij}.\delta(t_1 - t_2) \qquad [2.49]$$

Let us introduce a "parasite" x_p.

The solution for the harmonic oscillator equation:

$$\frac{d^2V}{dt^2} + \omega_0^2 V = 0 \qquad [2.50]$$

can be written as:

$$V_T(t) = V_0 \cdot \cos\left(\omega_0 t + \varphi\right) + k.x_p \qquad [2.51]$$

where x_p can be called a Langevin noise source.

After some tedious, but not very difficult calculations, the solution (we assume the frequency is constant) for x_p, to the first order, is:

$$V_S(t) = V_0 \cdot \cos(\omega_0 t) + \gamma[(\cos(3\omega_0 t) - \cos(\omega_0 t)) + 12\omega_0 t.\sin(\omega_0 t)] \qquad [2.52]$$

In this solution, we encounter a time-proportional term. The oscillator must diverge (in fact, the amplitude of the voltage must be limited by the power supply). We suggest that the damping response – see equation [2.51] – could be compensated by the term x_p explained in equation [2.52], especially for a big enough D_{-1}.

2.5.3. *Influence of digital blocks on analog blocks*

Stochastic behavior of digital block perturbations

The well-known SPICE or ELDO tools are able to model phenomena associated with substrates and digital spurious signals; but their applicability is overly optimistic for complex technology with very large digital blocks. Some noteworthy attempts have been made to model the generation of noise injected into the mass from a "SPICE" point of view and a process/device point of view, but it seems accurate for relatively small circuits and simple parasitic waveforms, moving through the substrate.

Today, CAD tools cannot be used, ab initio, to slow down paths that are not critical for timing, and thus reduce parasitic coupling mechanisms. However, a placement option exists for use during the placement/routing step to attain this objective; this option is called "recovery" and is used to reduce the size of logic gates that are oversized. Its first objective is therefore to reduce the surface area of digital portals and it also reduces the overall noise of a digital block by reducing all unnecessarily fast transitions. While this step may take a long time, it should be worthwhile including it in the digital back-end stream of the mixed-signal IC.

The most practical way to go further in the efficient management of digital noise of (very) large circuits seems to be statistical; Monte Carlo procedures seem the tool of choice, but this is time-intensive for large ASICs. Another way is to introduce the Markov chain theory.

Parasitic signal injections, and their possible interference, can come from physical phenomena, such as impact ionization (from CMOS cells or at P/N junctions). Interference, induced by the digital part – the subject of our investigation – can spread in different ways: power supply network, signal wires, package pins, links and obviously the silicon substrate. To put it briefly, in a Markov process: "the past has no influence on the future if the present is specified".

In this last section, the stochastic methodology is based on probabilistic modeling of digital switching activity. The ultimate goal is to calculate the power spectrum of interference noise (see the synoptic block diagram in Figure 2.37(a)). This spectrum should have both a discrete and a continuous component; the former is involved at a periodic rate, triggered by the digital circuit clock(s), the continuous component resulting from the stochastic nature of the events. Ultimately, the resulting spectrum will depend on the shape of the injected interference signals. First, we will describe the methodology, as generally as possible, that is relevant to accessing such a noise spectrum. The idea described in this section has already been proposed; in this book, we present its development, elaborate details and finally present, for the moment, simple examples for the practical application of the methodology we have implemented.

Methodology

Our topics of interest are digital transmission/propagation systems. For a general approach, consider noise sources specific to M (parasites), x1(t), x2(t), ..., xM(t), continuous signals in time, modulated by a coded message c[n]. This can be written in the general form:

$$x(t) = \sum_{n=-\infty}^{\infty} g^{c[n]}(t - nT + \varphi) \qquad [2.53]$$

This equation means that the c[n] "symbol", the switching event, emitted at time nT, assigns the waveform g, where T is the period of the digital clock. For now, we take Φ to zero (but Φ could be a stochastic variable: a possibility to model phase noise). Either s_n a stationary Markov chain, with a defined state: $S = \{\sigma_1, \sigma_2, ..., \sigma_I\}$ and c[n] = f(s_n), or a vector-valued stochastic process is obtained as a function without memory of s_n. We assume that the Markov chain s_n is ergodic (s_n has only one class of communicating states). No degeneration is involved in this process; a time average is equal to an overall average. Here, g is a matrix, which has the general form:

$$\mathbf{g}(t) = \begin{bmatrix} g_1(t,\sigma_1) & g_1(t,\sigma_2) & . \; . & g_1(t,\sigma_I) \\ . & . & . \; . & . \\ . & . & . \; . & . \\ . & . & . \; . & . \\ g_M(t,\sigma_1) & g_1(t,\sigma_2) & . \; . & g_M(t,\sigma_I) \end{bmatrix} \qquad [2.54]$$

c[n] can then be written as c[n] = S_n C, where C is a matrix $M \times I$, and where S_n = [S_n (1), S_n (2), ..., S_n (I)], defined by S_n (i) = 1 if s_n (i) = σ_i, and otherwise S_n (i) = 0.

The jth column of C can be written as:

$$[f_1(\sigma_1), f_1(\sigma_2), \ldots f_1(\sigma_n)] \qquad [2.55]$$

and the jth row of C is the value of c[n] when $s_n = \sigma_i$.

For our case of a start of demonstration of a digital transition (I = 4), for a single noise source, (M = 1), we can use the simple notation, g_{ij}, where indices ij are primary states from state i to state j, where i and j are 0 (low state) or 1 (high state). The digital transition signal is defined as a function $f : S \to C$, where $C \equiv \{00, 01, 11, 10\}^T$ (o$\{LL, LH, HH, HL\}^T$). Then, from s_n, we could establish a transition probability matrix (TPM) P of general term P(i,j) (=Pr{$s_{n+1}=s_j/s_n=s_i$}).

Then, the relevant noise vector, here the simple < $g_{00}(t), g_{01}(t), g_{10}(t), g_{11}(t)$, will be assigned to time t (for each noise source, if necessary).

The "superposition" of all associated parasitic waveforms (which can be stored in a library), g, from the switching gate, will yield the final modulated signal. In addition, c[n] must be modeled according to a Markov chain. The very purpose here is to calculate the spectral power density, denoted as Sy(f), at any relevant circuit node. This density will be represented in matrix format, with algorithm development based on concise and broadly applicable matrix formalism.

Dealing with interference noise means developing a second-order analysis of the process, assuming that it is a second-order stationary system. In practical terms, this involves computing the matrix correlation function of Rx(k) (for the noise source x) of vector g(t ; c), that is:

$R_x[k] = E\{g(t;c[n])^*.g(t,c[n+k])\}$ (*: conjugate transpose, for the general case). [2.56]

Consider a single cell; each cell will furthermore be considered as a current source or sink, which can be injected or absorbed at the power supply/ground/ substrate/package/board hotspot, which will be assumed to have linear electrical behavior. The state of the digital signal can be set as follows:

L → H (or: 0 → 1): with probability α

H → L (or: 1 → 0): with probability β,

(L and H mean high and low, respectively).

Figure 2.37(a) represents the state diagram of such a "machine". Suppose that the "input" i is itself a Markov chain with a transfer probability matrix:

$$\pi = \begin{bmatrix} 1-\alpha & \alpha \\ \beta & 1-\beta \end{bmatrix} = \left(\begin{bmatrix} \pi_{00} & \pi_{01} \\ \pi_{10} & \pi_{11} \end{bmatrix} \right) \quad [2.57]$$

where: $\sum_j \pi_{ij} = 1$, i (= 0 or 1): row number; j (= 0 or 1): column number.

$\pi_{01} = \Pr\{j=1, \text{if } i=0\} = \alpha$ and $\pi_{10} = \Pr\{j=0, \text{if } i=1\} = \beta$

So: $\pi_{00} = \Pr\{j=0, \text{if } i=0\} = 1-\alpha$, and $\pi_{11} = \Pr\{j=1, \text{if } i=1\} = 1-\beta$.

A simple digital signal that can pass from bottom to top with probability 1 – α and from top to bottom with a probability 1 – α at each clock step is modeled by four states (see Figure 2.36(b)); it has a state TPM representing the switching activity, expressed by a matrix P, $I \times I$:

$$P = \begin{bmatrix} 1-\alpha & \alpha & 0 & 0 \\ 0 & 0 & \beta & 1-\beta \\ 1-\alpha & \alpha & 0 & 0 \\ 0 & 0 & \beta & 1-\beta \end{bmatrix} \quad [2.58]$$

Our application can be based on communication theory (see deterministic finite-state sequential machine, etc.). It can be directly studied as spectral properties of coded messages. This is a direct application of the communication theory field, although it is concerned with the propagation and interference of parasitic signals. The general term of P can be understood as: Pij/kl = Pr {state transition, from i to j, at n+1 time/if transition k to l to n}. Note that although matrix P is reducible (some states do not communicate with others), this does not imply any disadvantage for algorithm development.

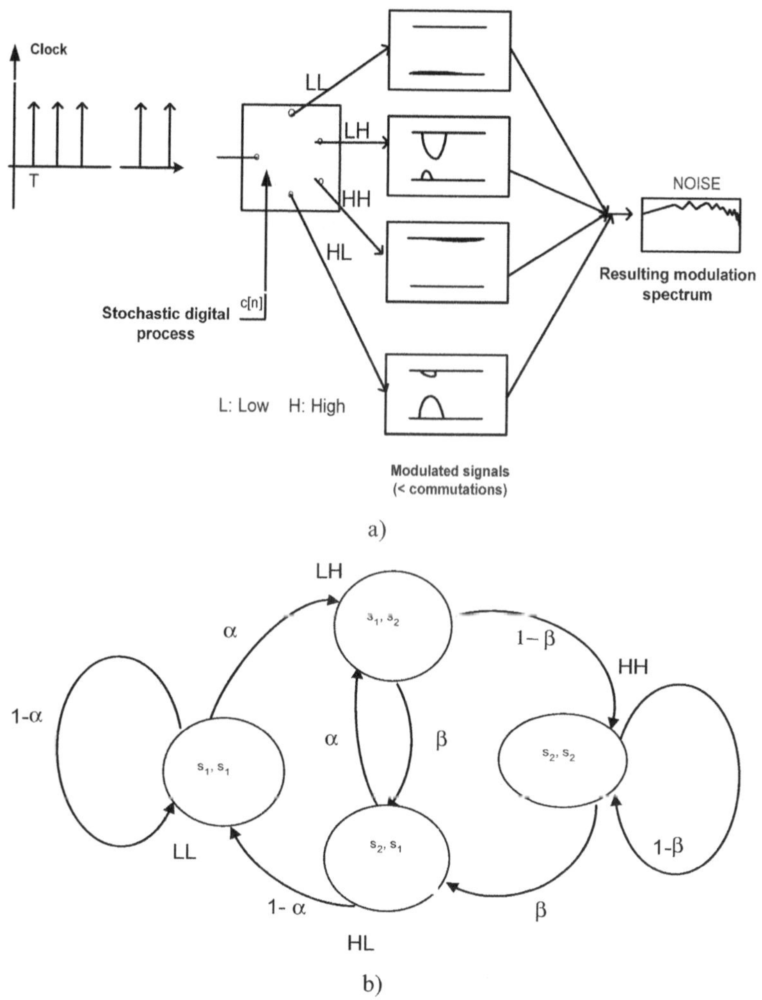

Figure 2.37. *(a) Functional diagram of the parasitic signal propagation chain. (b) 4-state sequential machine*

A somewhat cumbersome but not so difficult method to solve this problem is described as follows. As previously mentioned, the objective is to obtain the spectral density generated by digital block switches; this means finding the Fourier transform of the corresponding autocorrelation function (see second-order moment).

Spectral density w(f) can be broken down into two terms:

$$w(f) = w_d(f) + w_{cont}(f)$$

where f is the frequency.

The discrete component (W_d) has spectral lines or jumps at the clock frequency and its multiples (f =n/T), while the continuous part (w_{cont}) can be derived directly by considering z = exp(j*2πf) in the z-transform of its correlation function:

$$w(z) = \sum_{k=-\infty}^{\infty} R_{cont}[k]\ z^{-k} \quad (R_{cont} \text{ is explained below}) \qquad [2.59]$$

In addition, the discrete random process S can be characterized by an autocorrelation matrix:

$$R(k) \equiv E\{S_n^T(j)S_{n+k}(i)\} = \Pr\{s_{n+k} = \sigma_j, s_i = \sigma_n\}(T\text{: transposed}).$$

The entire correlation matrix (of the source codes c[n]) can be derived using the matrix algebra tool as follows:

$$R_c(k) = C^T R(k) C \qquad [2.60]$$

where R(k) is the joint probability matrix between random vectors Sn and Sn+k, with the codeword encoding c[n] = SnC, where C is a matrix to which each column assigns a specific parasitic waveform. From a general point of view, a source code b[n] is encoded in a sequence of c[n] codeword encoding; the “message” c[n] is generally a nonlinear transformation, namely c[n] = f(b[n]), and can take the general form (input-output relationship): $c[n] = g_0\, b_n + g_1\, b_n - 1 + \ldots + g_N\, b_{n-N}$; i.e. c[n] = f(b[n]) and can take the general form (input-output relationship):

$$c[n] = g_0\, b_n + g_1\, b_{n-1} + \cdots + g_N .\, b_{n-N}$$

where $b[n] = (b_n, b_{n-1}, \ldots, b_{n-N})$.

Note, in our simple starting example, N = 4, with (see above):

$$g_0 \equiv g_{00}, g_1 \equiv g_{01}, g_2 \equiv g_{10}, g_3 \equiv g_{11}$$

Finally, we explain the entire correlation matrix as follows:

$$R[k] = \begin{cases} \Delta P^k & k \geq 0 \\ \left(P^T\right)^{-k} \Delta & k < 0 \end{cases} \qquad [2.61]$$

P^k represents the *k*th step of the TPM.

Note that for k = 0, R(0) = Δ, and $E\{S_n(j)S_n(i)\} = p_i \cdot \delta_{ij}$, where δ is the Kronecker symbol with Δ = diag{p(1), p(2), …,p(I)}, which can be denoted as:

$$\Delta = \left[p_j^{\infty} \delta_{ij} \right] \qquad [2.62]$$

Elements of this diagonal matrix are stationary vector components (see infinity symbol), which are derived from the equation:

$$p = \Delta P \text{ (with } p = [p_1, p_2, p_3, p_4] \text{)} \qquad [2.62b]$$

R[k] can also be broken down into two "orthogonal" parts (see Vold's theorem): one continuous and one discrete:

$$R[k] = R_{cont}[k] + R_d[k] \qquad [2.63]$$

where:

$$R_{cont}[k] = \begin{cases} \Delta H^k & k > 0 \\ \left(H^T\right)^{-k} \Delta & k < 0 \\ \Delta - \sum_{r=0}^{h-1} K_r & k = 0 \end{cases} \qquad [2.64]$$

and a periodic component, with the period h, given by:

$$R_d[k] = \sum_{r=0}^{h-1} K_r.\exp(j2\pi rk / h) \quad [2.65]$$

where $R_d[k]$ is cyclic when h > 1.

We deal with the case of arbitrary h, but, without loss of generality, we do not explicitly represent the global matrix, with h > 1, which is itself constructed via a cyclic form of the master matrix P. We can highlight the discrete and continuous part of these matrices, by transforming them, by appropriate permutations of rows (and/or columns), so that only zeros appear in the main diagonal. K_r is the amplitude of jumps (in our example, induced by a clock), and we have:

$$K_r \equiv J_r{}^*p^T p J_r \quad [2.66]$$

(the asterisk denotes the transposed conjugate of the matrix, the Hermitian conjugate, for the general case). By using the matrix tool, K_r is directly linked to the eigenvalues of a modulus equal to one of the MTPs (p, the stationary probability vector of the MTP, is an eigenvector of P). In our problem, the solution gives the vector:

[α∗(1−β)/(α+β), (α∗β)/(α+β), (β∗α)/(α+β),β∗(1−α)/(α+β)] as a solution.

Matrix H is defined as follows:

$$H \equiv P - \sum_{r=0}^{h-1} \lambda_r J_r^* wpJ_r \quad [2.67]$$

where w = [111…11]': column vector.

This implies that H = P – P∞ (P∞ is the limit of P_k when k → ∞). Since H has all its modulus eigenvalues < 1 (by construction: P is divided into two parts, H being the continuous one); therefore, Rcont(k) is absolutely summable. Here, J_r has the general form:

$$J_r = \begin{bmatrix} \lambda_r^{-1} I_1 & 0 & -- & 0 \\ 0 & \lambda_r^{-2} I_1 & -- & 0 \\ ----- & ----- & --- & --- \\ 0 & 0 & -- & \lambda_r^{-h} I_h \end{bmatrix} \quad [2.68]$$

λr = exp(j2pr/h), where r = 0, 1, h – 1, has a modulus equal to 1.

In our example, there is only one modulus value: λ =1 (others are: 0 of order 2, and 2-α – β); in this case, we have:

$$R_d[k] = K_r \tag{2.69}$$

We substitute [2.65] in [2.60] and taking into account the fact that the limit of the matrix sum $M_i = \sum_{k=1}^{i} H^k z^{-k}$ converges into the matrix $M_\infty = zI - H$ (this result is obvious if we use the matrix formula: $\sum_{k=0}^{\infty} A^k = (I-A)^{-1}$, for any matrix A, provided that all its eigenvalues have a modulus less than unity). The eigenvalues λ_h of H are zero, or non-zero with a modulus less than 1 (λ_h = 0, 0, 2-α–β).

After some tedious and relatively easy calculations, the continuous part of w(z) (see equations [2.65] and [2.61]) can be written as:

$$w(z) = (z^{-1}I - P^T)^{-1} C\ (zI - P)^{-1}, \text{ where } C = \Delta - P^T \Delta\ P \tag{2.70}$$

z = exp(j*2πf) and I is the identity matrix.

As mentioned above, spectral density can be considered as the sum of two parts: continuous density and numerical density. The spectral density of the modulated signal x(t) can be formally written as follows:

$$S_x(f) = S_x^{\text{cont}}(f) + \sum_k \delta\left(f - f_k^d\right) \cdot S_{x,k}^d \tag{2.71}$$

The continuous spectral density of the modulated signal x(t) is given by:

$$S_x^{\text{cont}}(f) = \frac{1}{T} \cdot G^*(f) \cdot W_x(e^{j2\pi fT}) \cdot G^T(f) \tag{2.72}$$

where e G(f) is the Fourier transform of $g(t)$ and W_x (equation [2.19]) is the Fourier transform $R_x(\tau)$, where:

$$R_x(\tau) = \frac{1}{T} \cdot \sum_{k=-\infty}^{+\infty} \int_{-\infty}^{+\infty} g^*(t) R(k) g(t + \tau - kT) dt \tag{2.73}$$

Indeed, note that Rx(|) depends on the statistics of the Markov chain s_n. However, the noise signal x(t) has spectral lines (hops) at frequency multiples of f0 = 1/T, with an amplitude as follows:

$$ {}^{d}_{wx}(r) = C^{*}K_{r}C \quad , r = 0,1, \ldots, h-1 \qquad [2.74] $$

Thus, $w_x^d(r)$, with K_r defined by [equation [2.15]), becomes:

$$ w_x^d(r) = \left| \frac{1}{T} pJ_r G_r(f_0) \right|^2 \qquad [2.75] $$

(G is the Fourier transform of the matrix g(t)).

The P MTP has only one eigenvalue, α_0, equal to 1; all module 1 eigenvalues give Dirac distributions. In our case, only f_0 gives a train of Dirac distributions at the frequency 1/T and its harmonics.

Our problem is analogous to a particular modulation in digital communication. Then g(t:c[n]) can be written directly as c[n].f(t), f(t) being the basic waveform spurious vector. Finally, g(t) ≡ f(t).C.

Therefore, the continuous part of the spectrum becomes:

$$ w_x^{\mathrm{cont}}(f) = \frac{1}{T} \cdot |F(f)|^2 \cdot Wc(e^{j2\pi fT}), \text{ where } w_C(e^{j2\pi fT}) = C^{*} w(e^{j2\pi fT})C \qquad [2.76] $$

T the continuous part is: $S_x(f) = \frac{1}{T^2} \cdot |F(f)|^2 \, w_c^d$ [2.77]

Now, if we study the propagation of noise, we must consider the transfer function of the "propagation channel", denoted as T(f) (assuming network linearity). We obtain the noise density spectrum, considering this parasitic transfer function of the network, T.

The final power spectrum will be:

$$ S_y(f) = T(f) \cdot S_x(f) \cdot T^{*}(f) \qquad [2.78] $$

where S_x(f): PSD of noise sources obtained using modulation theory and $T(j2\pi f)$: (usually) multi-input system (noise sources) and linear multi-output (number of hotspots).

A simple application: rebound of the mass via the substrate

To illustrate the algorithm explained above, we study the case of signals injected into the analog electrical network by a digital switching cell. In this part, we use a simple model, while the ultimate goal will be to simulate the entire substrate voltage. For the time being, we will consider power bounces as the single source of substrate noise (knowing that this method can easily be generalized). To illustrate, we present some noise measurements, concerning a 90 nm CMOS technology (see Figure 2.37); in fact, a test chip was made in a 0.35 μm BiCMOS process (with five metal layers and a highly resistive substrate, from STMicroelectronics). In a CMOS inverter, the NMOS source transmits bounces to ground and the PMOS source transmits power supply fluctuations.

Power and ground fluctuations are generated by large transient current peaks drawn by the simultaneous switching of multiple digital gates. These pulses as well as the parasitic elements of the housing, PCB, padring and other parts of the power and ground networks can introduce resonance channels and therefore a power and ground resonance. Power supply oscillations are directly coupled to the substrate by a P+ contact. These perturbations affect analog blocks in the same way as they were generated. The coupling with the substrate can be resistive through substrate bias contacts, capacitive through interconnects and depletion capacitors, nonlinear through the MOS body effect mechanism, for example, or other effects related to each basic analog component. Variations in the operation of each component can cause serious signal integrity problems in the overall analog system.

This should allow us to characterize the substrate noise generated by the digital part of this type of small mixed signal integrated circuit. The package used is a QFN (Quad Flat No leads) 5 × 5 with 28 pins. It was chosen for its very short connecting wires and rear connection. This type of package is currently offered for radio frequency and low-noise applications. On the matrix, four inverter arrays are placed to generate noise in the resistive substrate. They can be controlled, with many configurations, by a programmable digital control. The inverter network control is isolated from the substrate by a P+ protection ring connected to a dedicated ground point. This substrate protection should limit the impact of digital control switching events on the measured disturbances.

All inverters can switch in a very wide frequency range from 0 to more than 500 MHz, depending on the external clock frequency. On the same matrix, there is a 4.5 GHz voltage controlled oscillator (VCO), with a dedicated power supply network. This analog device must be perturbed by digital switching noise. Some loose contacts, all over the array, allow for substrate voltage measurements and external noise injection to disrupt the VCO. Four of them are directly connected to

50 Ω measurement lines by bonding wires and on-chip metal lines. Other substrate contacts will be useful for measurements on silicon with microprobe installations.

Numerous impedance measurements (21 substrate contacts), with micro-probes, can be made between the different substrate contacts that are placed on the test chip. In fact, the main frequency of perturbation seems to be the resonance of the RLC path composed of the inductance and resistance of the bonding wire and the capacitance of the core of the digital part of the circuit; in other words, the power (and ground) lines must be modeled efficiently to predict voltage drops. Thanks to the ICEM, emission levels can be predicted in conducted mode on a chip as well as in the application board by means of SPICE-based analog simulations.

Thus, using this model to predict power bounces, and therefore substrate voltage, could be of interest. We can describe each part of the power lines with this model. Each part is modeled by an RLC (Figure 2.38). We can very easily deduce the following relationship:

$$V_{core} = V_{\sup ply} - \frac{L_{\sup} \cdot j\omega + R_{\sup}}{R_{\sup} C_{core} \cdot j\omega - L_{\sup} C_{core} \omega^2 + 1} \cdot I_{act}(j\omega) \qquad [2.79]$$

Values used for parasitic components that model the power grid (considered a single voltage source) are based on measurements and extractions. Typically, we write: R = 0.5 Ω, L = 0.2 nH, C = 0.2 nF. Power fluctuations do not depend on the topography of the design. In our case, and for future developments, for a digital block, the switching activity current will be applied to one point: the central point of this block. As an example, to validate the methodology and give an overview of some aspects of the problem, we consider a gate with only the state transition LH and HL (since state transitions LL and HH involve a very weak parasitic signal); the shape of the parasitic signal LH and HL will be assumed to be Gaussian with respect to time (this is not far from reality, and considering moments greater than 2, for the shape of the signal, is not useful, for the moment, to validate our algorithm; moreover, a library of parasitic signals is the best solution, for an industrial circuit).

LH: delay: 20 ns; standard deviation: 0.001 ns. $V_{peak} = 0.5$ V.

HL: delay: 60 ns; standard deviation: 0.01 ns. $V_{peak} = -0.15$ V.

To illustrate the algorithm explained above, we study the case of signals injected into the analog electrical network by a digital switching cell.

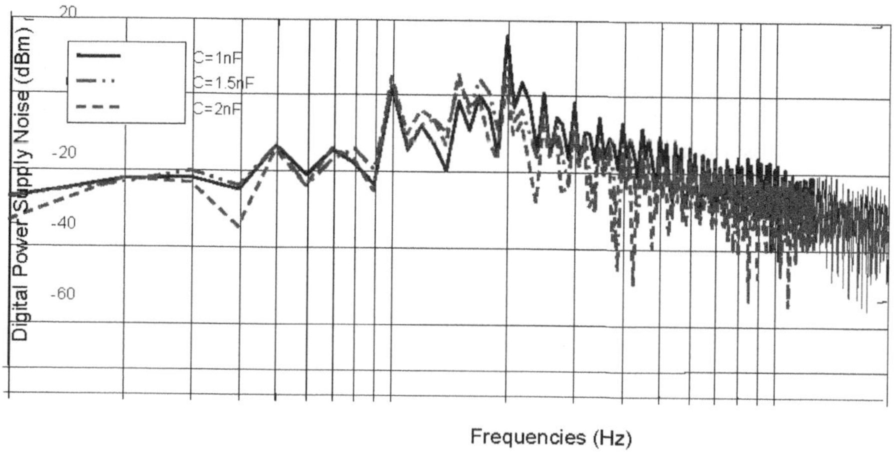

Figure 2.38. *Typical power supply noise measurements, with different basic capabilities. For a color version of this figure, see www.iste.co.uk/gontrand/analog2.zip*

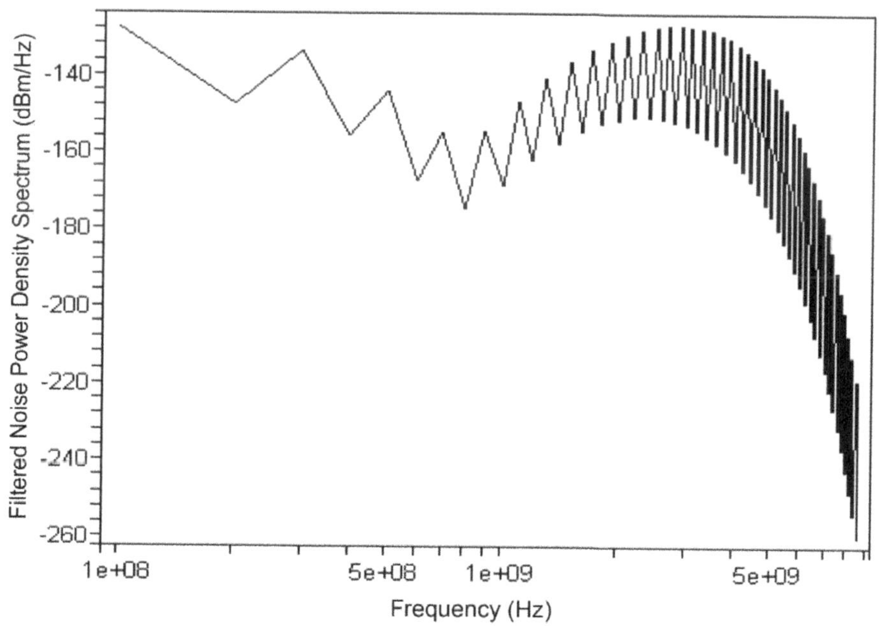

Figure 2.39. *Overall noise spectrum/typical calculation of a gate disturbing a power supply network*

Figure 2.39 shows the first simulations, which can be produced by a switching gate through an RLC parasitic type network from this Markov stochastic algorithm; it represents the continuous and discrete parts of the power spectrum of the parasitic waveforms injected into the V_{DD} line. Periodic behavior in this spectrum can be verified corresponding to the clock frequency.

3

From 2D to 3D: Opportunities and Challenges

Three-dimensional (3D) integration is perhaps the most promising technological solution to keep up with the level of integration dictated by the famous Moore's law, and not just for a few niches. From the beginning of the millennium, it has led to important research work. It allows different circuits and devices to be superimposed in a single package. Its main advantage is to allow a combination of heterogeneous and highly specialized technologies for the constitution of a complete system, while preserving a very high level of performance due, in particular, to very short connections between these different circuits or strata.

One of the objectives of this book is to provide models of substrates coupled with their internal connections and/or contacts through/in the substrate with several degrees of finesse/precision to allow high-level designers to manage and especially to optimize the partitioning between different strata. This modeling involves developing several views at different abstraction levels: from the physical model to the "high-level" model.

3.1. Introduction

The electronics industry has been the largest industry since 1996. A significant contributor to the industry's prominence over the past 50 years has been Gordon Moore's 1965 proposition of doubling the number of transistors on a chip every 24 months, famously known as Moore's law. This principle has exerted a remarkable influence over the microelectronics sector for the last five decades, primarily enhancing scaling and integration through two-dimensional (2D) lithography of various functions on a single chip. These integrated technologies serve the objectives of systems on chip (SOC) and system in packaging (SIP).

As a result of the requirement for higher clock frequencies and lower energy consumption, nowadays, these traditional 2D planar technologies are starting to face technical and financial challenges: physical limitations, processing complexity, manufacturing costs, etc. This is a strong increase in the "quantitative" and "qualitative" on certain numbers of characteristics, necessary for current and future multimedia or mobile applications; the complexity of the design is increased exponentially. Scaling the technology as well as integrating disparate technologies into a single chip means that device performance continues to outpace interconnect domains and encapsulation capabilities and therefore there are many technical challenges, including power management, intra-level heat dissipation, and chip cross-communications. Interconnection line delays are also becoming an obstacle in ultra large system integration (ULSI) 2D systems with respect to power dissipation, transistor delays, etc. Non-scale technology approaches are now being investigated in order to follow or to achieve more than Moore's law.

3.2. 3D integration

The new 3D integration technology with smaller form factors, higher integration density and the ability to make mixed technology chips seems to be the most promising option. It is considered to be an effective way to improve performance over conventional 2D schemes. There are numerous potential benefits to 3D integration such as improved performance and combinations of flexible heterogeneous systems (CMOS logic, RF analog function and memories). Vertical IC Si stacks allow for significant shortening of interconnect line arrays (Figure 3.1), thereby decreasing interconnect line delays, which become a major obstacle in V(U)LSI 2D (V: Very) systems compared to transistor switching delays and energy dissipation, while increasing integration density, leading to smaller form factors and reduced manufacturing costs.

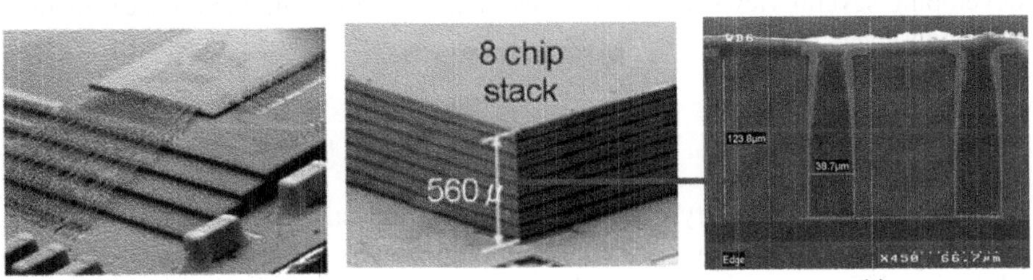

(a) (b) (c)

Figure 3.1. *(a) Stacked chips with wire bonding. (b) 3D TSV: stack structure (Source: SAMSUNG). (c) Cross sectional view of the 3D TSV stack structure (source: CE3.LETI)*

Currently, because of the control of micro- and nanotechnologies, 3D-integrated systems can be obtained by stacking 2D-integrated circuits vertically using “wire-bonding” (Figure 3.2). But wire-bonding does not allow for the connection of chip input/output pads located at their perimeter. In addition, wire-bonding has drawbacks in terms of surface delays and propagation as a function of clock frequency and wire lengths (Figure 3.1(a)). Generally speaking, 3D integration consists of 3D IC packaging, 3D IC integration and 3D Si integration.

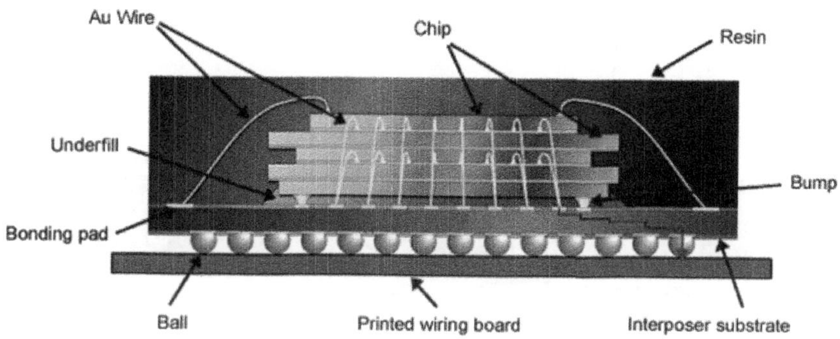

Figure 3.2. *Example of SIPO structure (stacked 5-chip SIP technology) (source: Renesas). For a color version of this figure, see www.iste.co.uk/gontrand/analog2.zip*

3D technology faces many technological challenges, such as the realization of vertical interconnections that ensure signal transmission, the link with functional channel alignment and the thinned substrate.

Through Silicon Via (TSV) came at the right time and made 3D circuits more feasible. In fact, William Shockley had “intuitions” about TSVs more than 50 years ago but had not planned for 3D Si/IC integration at that time. Now, they play an important role in this technology; it separates the 3D IC “packaging” from 3D IC/Si integration, while the latter two use TSVs, but not the 3D IC packaging. There are also differences between 3D IC and 3D Si integrations: 3D IC integration stacks chips have bumps, while 3D Si integration stacks slices do not have these bumps.

An example of a 3D TSV stack with three circuits is shown in Figure 3.3. Electrical connections are ensured by new pre- or post-processing structures: metal redistribution layers (ReDistribution Layer (RDL)), copper pillars, and TSVs. Rear redistribution lines and a different density of TSVs allow for construction of dense and complex 3D interconnection structures from the mechanical and electrical point of view, connecting various stacked conventional circuits. RDL is used to distribute energy and signals at high speeds on top or bottom surfaces of a die and TSVs are a key technology for 3D integration, propagating signals through silicon layers in systems. Solder balls are placed at the back of the layers of the 3D systems to ensure the connection with their environment.

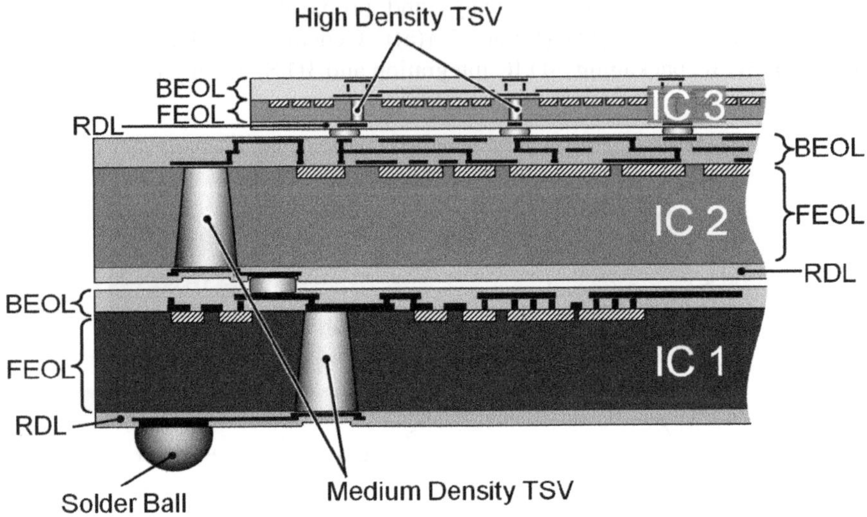

Figure 3.3. *Example of a vertical/3D stack with three ICs*

3D integration also allows for the integration of different types of chips and devices in a single package (Figure 3.4), or a compact subsystem offering maximum gain of highly specialized and heterogeneous technologies. However, in order to take full advantage of "3D IC", the decision must come upstream in the architecture planning process rather than as a packaging decision, after the circuit design is finalized. This requires us to take into account the 3D design space, from the beginning of the system design, in order to distribute its different strongly entangled parts.

The 3D TSV device gives a clear way of developing the electronics industry. These 3D integrated circuits, where interconnections are taken at the global or partial level in the chip, allow better integration, high bandwidth and electrical performance by reducing signal delays, with a smaller form factor and fewer pins, and overall better compactness. In this type of technology, the most common solutions then use stacked memory devices as well as sensors, in particular CMOS coupled with DSPs (digital signal processing). Future applications aim to package multi-core processors, hierarchical caches and other incompatible technologies into current heterogeneous integration solutions. Figure 3.5 shows a 3D-IC/TSV roadmap.

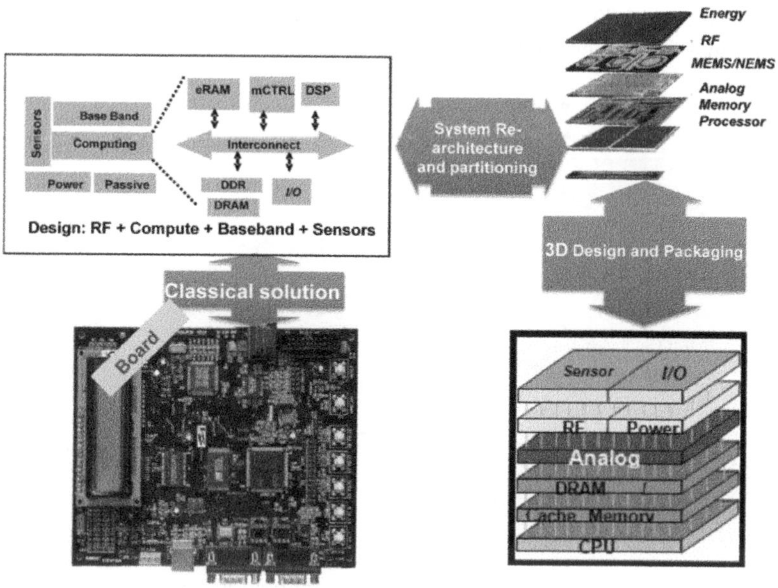

Figure 3.4. *Classic 2D solution versus 3D concept. For a color version of this figure, see www.iste.co.uk/gontrand/analog2.zip*

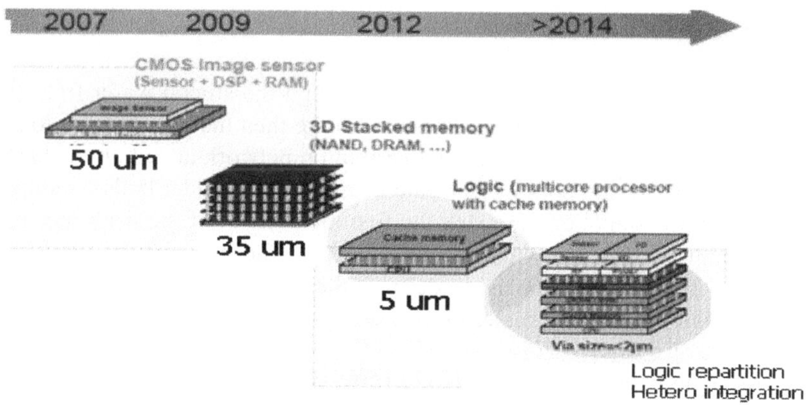

Figure 3.5. *Integration trend (STMicroelectronics/LETI). For a color version of this figure, see www.iste.co.uk/gontrand/analog2.zip*

CMOS image sensors using a "last via (via last)" approach, with diameters of about 50 µm and similar thicknesses of silicon, have already been introduced on the market. Figure 3.6 shows a 3D image processor that uses TSVs to incorporate an image sensor, an A/D converter, the frame memory and the reconfigurable image processor in a four-layer stacked configuration.

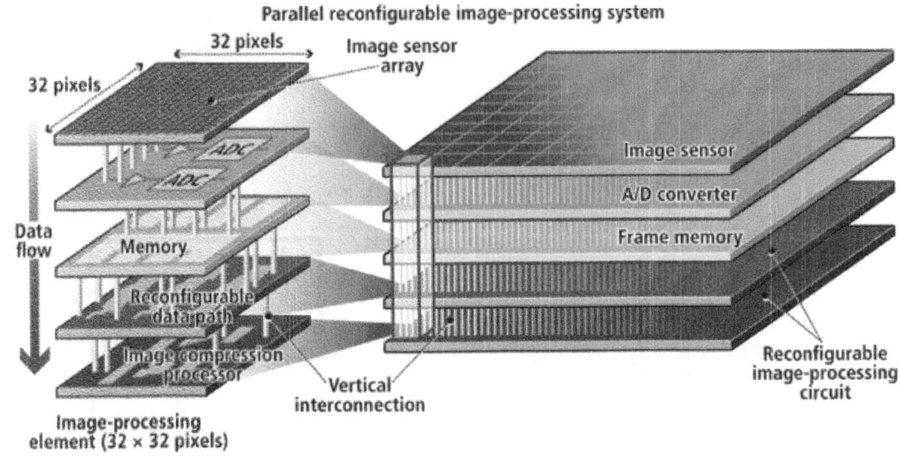

Figure 3.6. *A 3D image processor uses TSV technology that incorporates an image sensor, A/D converter, raster memory and reconfigurable image processor in a stacked four-layer configuration (source: vision-systems). For a color version of this figure, see www.iste.co.uk/gontrand/analog2.zip*

A first technique called "TSV via last" (which has already been mentioned) consists, after having produced the active devices on the front face of the silicon wafer, of thinning it and then drilling holes through the silicon wafer from the rear face to the metals of the front face. These holes are then metallized so as to connect the connections of the front face to a redistribution network added on the back face, allowing direct soldering on the electronic board (e.g. by solder balls – bumps). The difficulty of this technique is to reach the first metal layer of the front face reliably, without having to open too wide orifices, which would penalize the dimensions of the chip. The dimensions of holes, made from the rear face, depend on the technique used and the thickness of the plate.

Another technique also called "BEOL (post Back End Of Line) via first" can be applied to plates at the end of active component manufacture, but before plate thinning. It consists of etching and metallizing TSVs on predefined regions of the chip. Metallized TSVs are connected, on the front side, to the metal layers of the chip by means of a redistribution layer. Chip thinning then releases the metallized TSVs, which will be connected, on the rear face side, either directly or by redistribution metallization, to a metal layer (electrolytic copper type), suitable for stacking with soldering.

The most dynamic industrial players on the subject are as follows: IBM and Intel (massive integration of processor cores), Samsung (3D integration of memories) and Tezzaron. In France, STMicroelectronics, NXP and CE3-LETI have launched many programs to master 3D integration (Figures 3.7 and 3.8). Typically, French players work on vias in low density (diameter ~50 µm, form factor ~1–2) and vias in high density (diameter ~3 µm, form factor ~4–10). The first application is of the imager type, with products already existing at ST integrating the CMOS sensor on a digital chip.

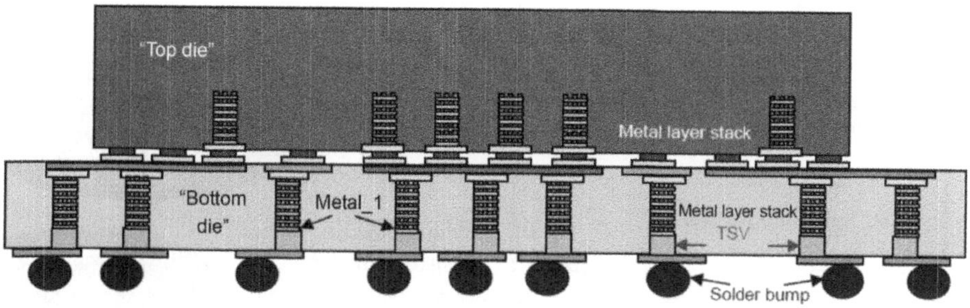

Figure 3.7. *Example of use of through via (drawing not to scale) (STMicroelectronics/LETI). For a color version of this figure, see www.iste.co.uk/gontrand/analog2.zip*

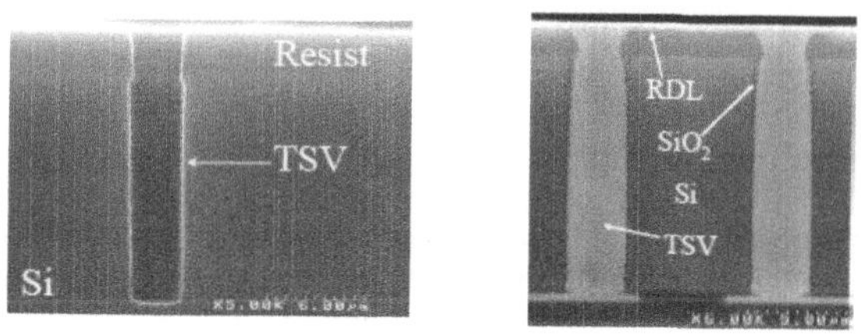

Figure 3.8. *Example of an embodiment of a through via (LETI)*

Application to the extraction of parasitic substrate elements

In the case of high frequencies, TSVs become a major source of substrate noise, which threatens the performance of neighboring devices and increases power consumption. So, in the case of this high frequency, in order to prevent the entire system (die, with its "packaging") against surges, bounces or parasitic oscillations, it is important to understand the characteristics of this substrate coupling. The loss of circuit performance caused by the silicon substrate can be very strong. Given the radio frequency (RF), the substrate could have, at certain nodes, an "RLC" behavior (essentially an "RC" for low frequencies).

As already mentioned, the substrate parasite extractor implemented in MATLAB as well as in Java for easy portability uses the Green's functions in order to generate an RC mesh of this substrate. The substrate is considered as a superposition of conductive layers of different resistivities (related to doping) and different dielectric permittivities. From a list of contacts, a priori rectangular in shape, the tool "returns" a matrix of resistors, or capacitors. Figure 3.9 summarizes the use of our IT tool.

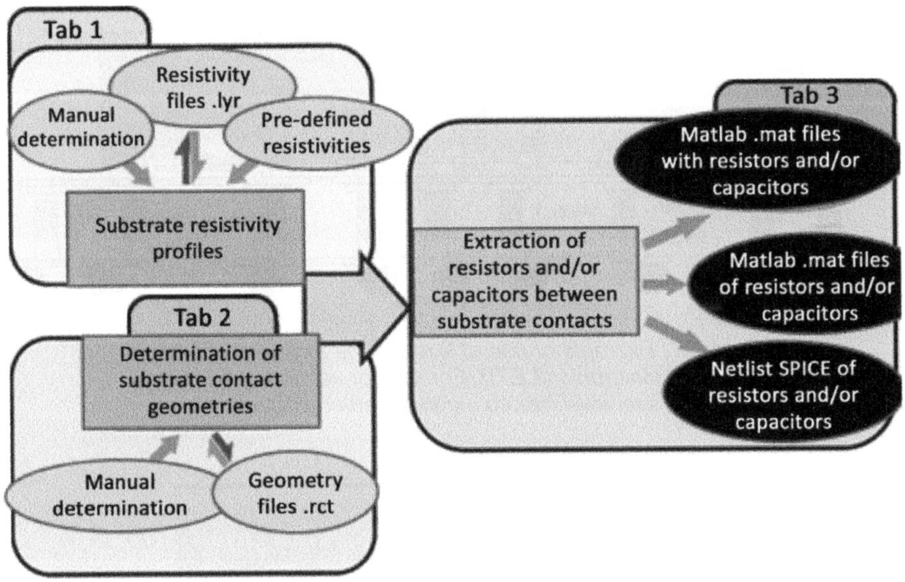

Figure 3.9. *Computer tool for extracting resistances and substrate capacitances. For a color version of this figure, see www.iste.co.uk/gontrand/analog2.zip*

In the case of substrate extraction by the Green method (of cores), the substrate is considered to be a superposition of homogeneous dielectric and/or conductive layers.

Figure 3.15 shows an integrated circuit design flow. As can be seen in the graph, between the layout design and the circuit simulation at the "component" level, there is a step called "pest extraction". The task of this step is to build a model of capacitors, inductors, resistors and other components for studying electromagnetic coupling; the entire circuit can then be accurately simulated. Of the three parameters – capacitance, resistance and inductance – capacitance is often the most prohibitive, as it severely affects circuit delays, power consumption and signal integrity. Simultaneously, the high-frequency circuit necessitates careful consideration of inductance effects, thermal and mechanical stress phenomena. In circuits, there is a notable influence of eddy currents in RF.

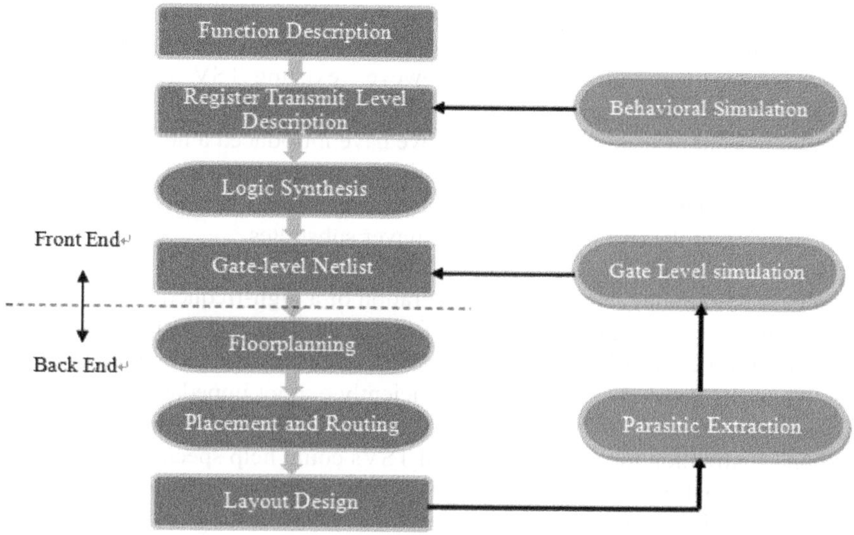

Figure 3.10. *A typical IC design technology process*

Gharpurey carried out pioneering work in modeling and analysis of substrate coupling in integrated circuits (ICs). The Green function must further be transformed into a numerically stable form, a priori suitable for "solving" the impedance matrix for an arbitrary three-dimensional arrangement of conductive contacts placed anywhere in or on the substrate. More recent algorithms present an improvement over the quasi-3D approach used in solvers also based on Green's functions and takes into consideration the 3D conductor flanks, namely, contact thickness.

3.2.1. *3D impedance extraction*

Efforts to expedite circuit design, while preserving accuracy and dependability, have led to the search for more efficient tools due to certain numerical method inefficiencies within the substrate model. The tool 1 is required to analyze topography, gauge the circuit's electrical behavior and compute the substrate's impedance matrix along with its coupling to connectors. Research programs focused on electrical, thermal and mechanical stress aspects have emerged in recent years. Although some approaches, like substrate extraction (employing the block element modifier (BEM) method for reusable component and code sharing in front-end development), have been proposed, they have been limited to analyzing coupling between surface contacts, excluding 3D-TSV structures. The examination of parasitic coupling capacity encompasses various configurations, ranging from the conventional placement at the bottom or on the side, depending on whether the TSV is encircled by interconnecting wires. However, existing TSV models simplify by omitting TSV inductance and oxide capacitance, assuming a high resistivity substrate. In light of these considerations, we have introduced a novel matrix method (transmission line matrix (TLM)) rooted in the well-established transmission line concept. This method is designed to extract impedance between contacts and/or TSVs, catering to the complexities of multilayer substrates.

As described above, high-density integration of a high-frequency system induces consequent noise couplings; this is one of the most important design considerations due to its significant influence on the performance of integrated circuits. The main objective of 3D substrate analysis is to efficiently extract impedance parameters (Z) between contacts and TSVs, which are on or in the silicon substrate. An effective impedance extraction tool for contacts and TSVs could help speed up the design and optimize the final layout. According to Ted Vucurevich, of all the EDA challenges for 3D SIC design, tools and methodologies for 3D IC testing are considered to be the "no. 1 challenge".

As with planar technologies, 3D interconnections can be understood via equivalent electrical models, such as "RLCG3" with a Π or T network. Often, a simple compact model is built for modeling the substrate as a simple node (Figure 3.11(a)). However, this assumption is only viable when the substrate is highly conductive, in the low and medium frequency ranges, and is not suitable for multilayer substrates. At high frequencies, substrate fittings can become a source of noise, which can lead to system errors and malfunctions, delays in signal adjustment, low reliability, and short system life; therefore, substrate effects can be accounted for by modeling via an "RLCG" network (see, for example, Figure 3.11(b)). This is why a robust method of associated substrate extraction is needed.

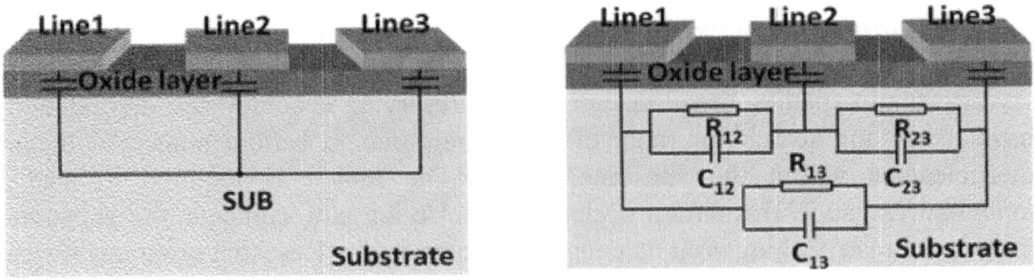

Figure 3.11. *(a) Front view: medium-density TSVs and modeling including current path (b) compact 'RLCG' modeling of the signal line and a coplanar waveguide ground line. For a color version of this figure, see www.iste.co.uk/gontrand/analog2.zip*

As a general rule, Z parameters can be defined as:

$$Z_{mn} = \frac{V_m}{I_n}\bigg|_{I_{k \neq n} = 0} \quad [3.1]$$

In practice, by injecting a unitary excitation current at a point n, the resulting potential at the point m is calculated; we can then obtain impedance Zmn between contact m and contact n directly.

If we can get the resulting potential distribution of the substrate caused by the unit current, we could get Z parameters by equation [3.1].

Under quasi-static conditions, the potential on the substrate satisfies Laplace's equation:

$$\nabla^2 \varphi(x, y, z) = 0 \quad [3.2]$$

It can also be written as:

$$\left(\frac{\partial^2}{\partial x^2} + \frac{\partial^2}{\partial y^2} + \frac{\partial^2}{\partial z^2} \right) \varphi(x, y, z) = 0 \quad [3.3]$$

After a spatial 2D fast Fourier transform (FFT), in x, y directions, we transform the spatial coordinate domain into a frequency domain, widely used method in digital image processing.

Thus, the above equation is transformed into:

$$\frac{\partial^2 \Phi}{\partial z^2} - k_x^2 \Phi - k_y^2 \Phi = 0 \quad [3.4]$$

where Φ is the potential in the "frequency" domain and k_x, k_y are spatial-frequency variables (wave vectors).

After a few transformations, we get:

$$\begin{cases} \dfrac{\partial \Phi}{\partial z} = -\dfrac{J_z}{(\sigma + j\omega\varepsilon)} \\ \dfrac{\partial J_z}{\partial z} = -(\sigma + j\omega\varepsilon)\left(k_x{}^2 + k_y{}^2\right)\Phi \end{cases} \quad [3.5]$$

These equations indicate the relationship between the current density in the z-direction and the potential distribution in the spatial-frequency domain.

Based on the TLM method, in two-port networks, voltage u (z, t) and current i (z, t) can be represented as follows in two-port networks:

$$\begin{cases} \dfrac{dU(z)}{dz} = -(R_0 + j\omega L_0)\cdot I(z) = -Z_1 I(z) \\ \dfrac{dI(z)}{dz} = -(G_0 + j\omega C_0)\cdot U(z) = -Y_1 U(z) \end{cases} \quad [3.6]$$

Comparing equations [3.5] and [3.6], we find that they are formally identical to each other.

Line voltage U in the "TLM" corresponds to the transformed potential distribution Φ, and line current I corresponds to component z of the current density in the transformed domain.

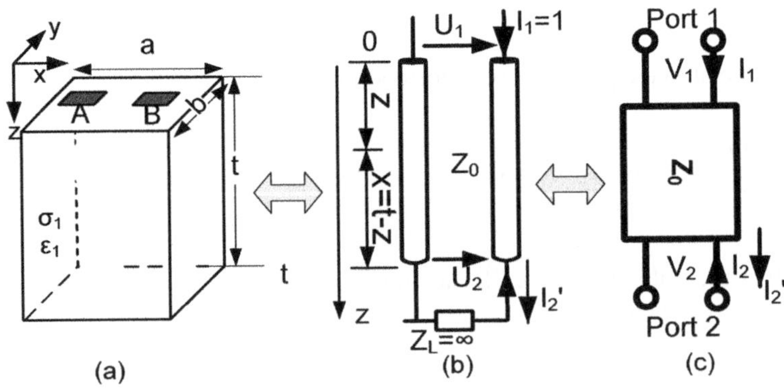

Figure 3.12. *(a) Two contacts provided on the top surface of a single-layer substrate. (b) Its equivalent TLM transmission line model. (c) Equivalent two-port network model. Since the current flowing on the bottom surface (a) is equal to zero, the load impedance (ZL) at (b) is positive "infinite"*

For a single-layer substrate model, the relationship between the substrate model and its equivalent transmission line is shown in Figure 3.17 (note the current direction of I_2', exiting the port: denoted as I_2' distinguish it from the injected current I_2).

A single-layer substrate could be modeled as a two-port network; a multilayer system as shown in Figure 3.12(a) can be modeled as the cascade connection of two-port networks (Figure 3.12(b)).

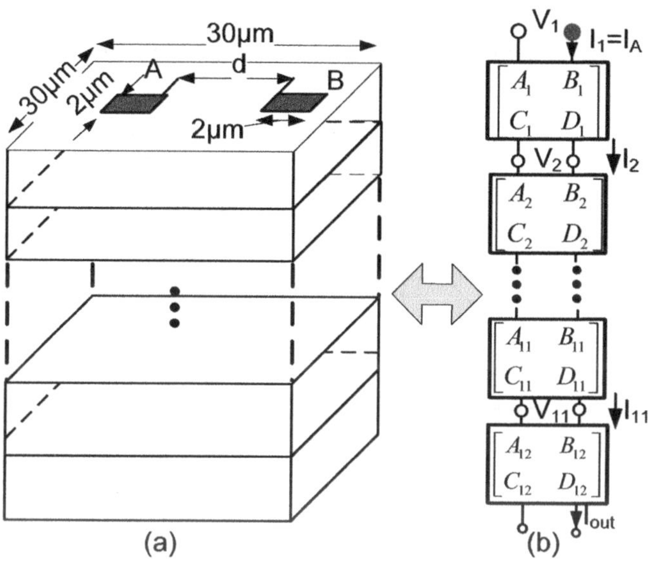

Figure 3.12bis. *Multilayers with surface contacts to be tested (a) and corresponding TLM model (b). The size of the substrate is 30 µm in the X, Y directions. Two square contacts of 2 µm × 2 µm are located on the upper surface of this multi-layer substrate with a separation distance d*

Z parameters can be converted from ABCD parameters (see chain matrix) of equation [3.7]:

$$Z = \begin{bmatrix} z_{11} & z_{12} \\ z_{21} & z_{22} \end{bmatrix} = \begin{bmatrix} \frac{\mathrm{A}}{\mathrm{C}} & \frac{\det(ABCD)}{\mathrm{C}} \\ \frac{1}{\mathrm{C}} & \frac{\mathrm{D}}{\mathrm{C}} \end{bmatrix} \qquad [3.7]$$

Then, Z_{11} used as Z_{in} can be obtained from equation [3.1]. We can then access the Z parameter.

Main strength of "3D-IE" (Impedance Extraction) is that it is well dedicated to integrated contacts. Two different cases will be discussed: integrated partial contacts and all buried contacts.

Figure 3.13 shows an integrated contact in a three-layer substrate. It is compared with Figure 3.12. The boundary condition is the same, namely, $I_{out} = 0$ and $I_1 = 1$. Given that contact B is not located on the upper surface, we need to know voltage V_3 corresponding to 3 (Figure 3.13), not V_1 like the model in Figure 3.12.

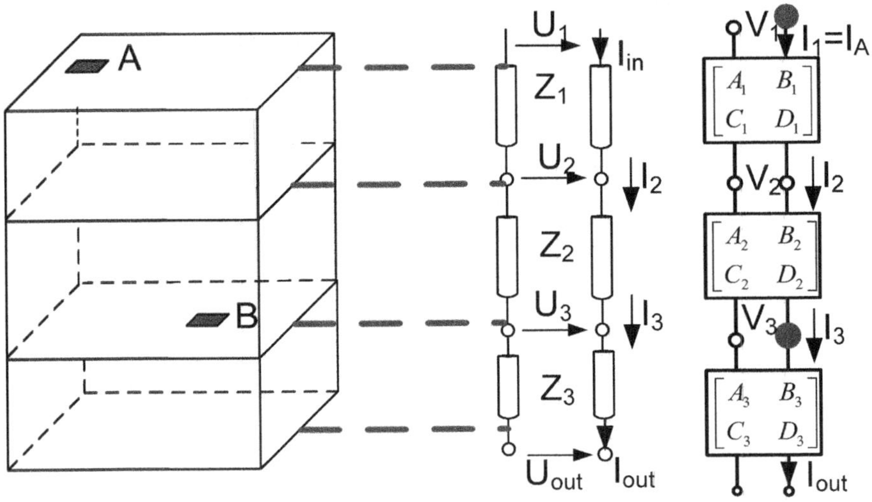

Figure 3.13. *Template for some built-in contacts*

While we get ABCD parameters for each layer, V_3 can be obtained easily under the condition of $I_{out} = 0$ and $I_1 = 1$. Then, Z_{31} is calculated in V_3 and I_1.

Z_{31} is used to calculate the potential in the spatial-frequency domain: an inverse 2D spatial Fourier transform is applied to obtain the real potential distribution (φ_z) of the xy plane on which contact B is defined. Then, equation [3.1] is applied to calculate Z_{BA}. Other Z parameter elements can be similarly obtained.

If all contacts are buried, the substrate can be considered as a paralleling of two separate parts, as shown in Figure 3.14.

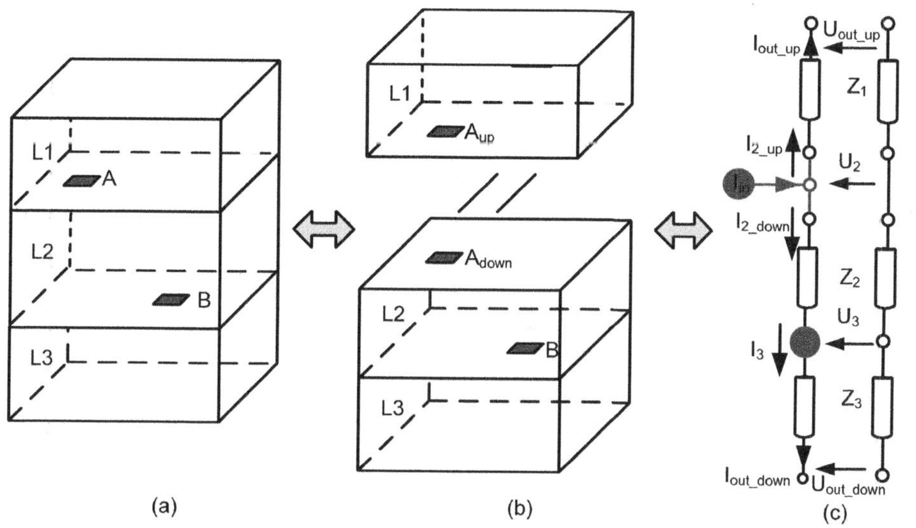

Figure 3.14. *Template for all integrated contacts*

The current is injected at contact A, so that the substrate is divided into two parts, from the layer. Figure 3.20(c) shows the equivalent TLM model. I_{out_up} and I_{out_down} are all equal to zero due to boundary conditions. I_{in} is a unit current ($I_{in} = 1$). In order to get I_{Z32}, we need to know voltage V_3. For the TLM substrate model of the upper half, assume that input impedance is Z_{in_up} and Z_{in_down} at the lower half level due to paralleling (see current divider):

$$I_{2\_down} = I_{in} \frac{Z_{in\_up}}{Z_{in\_up} + Z_{in\_down}} \qquad [3.8]$$

Z_{in_up} and Z_{in_down} impedances can be calculated via ABCD parameters of each layer as shown. As I_{2_down} is calculated, the model will be the same as shown in Figure 3.14. The following steps are exactly the same with partial contacts of the integrated model.

The potential of each layer can be calculated using this method. Figure 3.15 shows the potential of a substrate with two contacts in layer 6 and the other in layer 9.

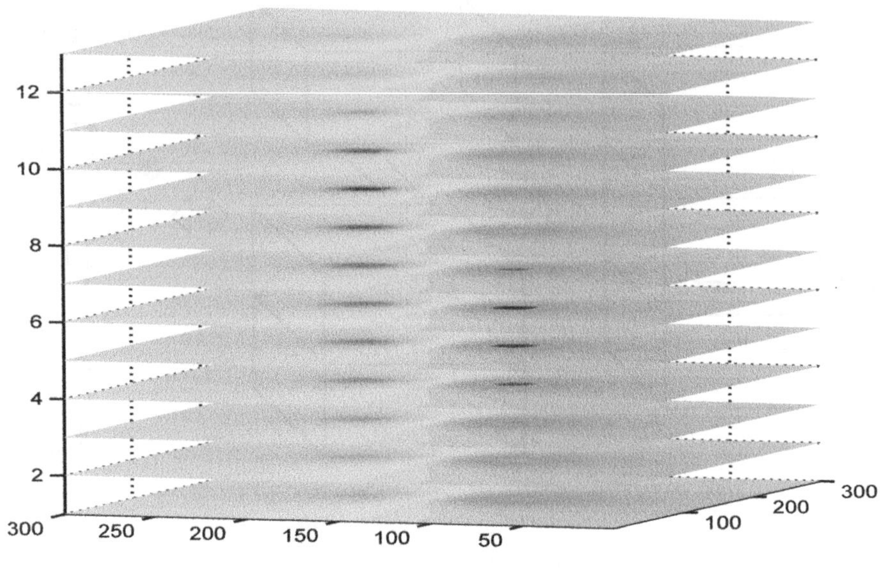

Figure 3.15. *Potential of each layer. For a color version of this figure, see www.iste.co.uk/gontrand/analog2.zip*

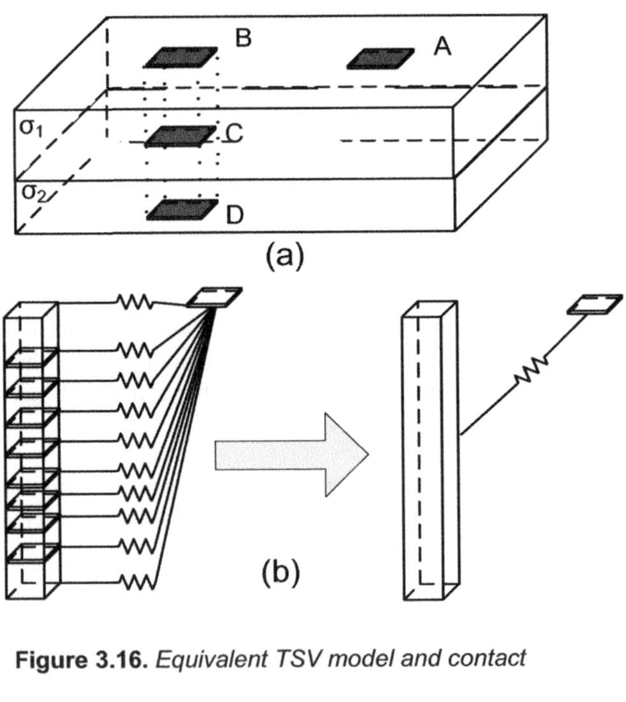

Figure 3.16. *Equivalent TSV model and contact*

Thanks to our analysis above, we can calculate the impedance between contacts on arbitrary locations and arbitrary layers. Thus, a special structure is considered in order to calculate the impedance involving TSVs, as shown in Figure 3.16(a).

Impedances between AB AC and AD can be calculated using the above algorithm. Next, we combine the coupling between AB AC and AD to obtain a coupling between contact A and contact "BCD". In fact, the "BCD contact", the combination of B, C, and D, is the TSV (here, we assume that the TSV is not surrounded by an oxide coating, TSV with oxide coating will be discussed later). A diagram is shown in Figure 3.16(b).

The TSV can be divided into small sections and each part examined as a contact; the above process can then be used to obtain the impedance between each small TSV part and contacts. Finally, all impedances are combined to obtain the final impedance between the TSV and the contact, as shown in Figure 3.22(b).

Using a similar algorithm, the impedance between two TSVs can be calculated by dividing both TSVs into small pieces. After obtaining the impedance between each of the "pieces" of the two TSVs, the total impedance between 2 TSVs can be obtained by combining all impedances.

Thus, the general process of the algorithm is shown in Figure 3.17.

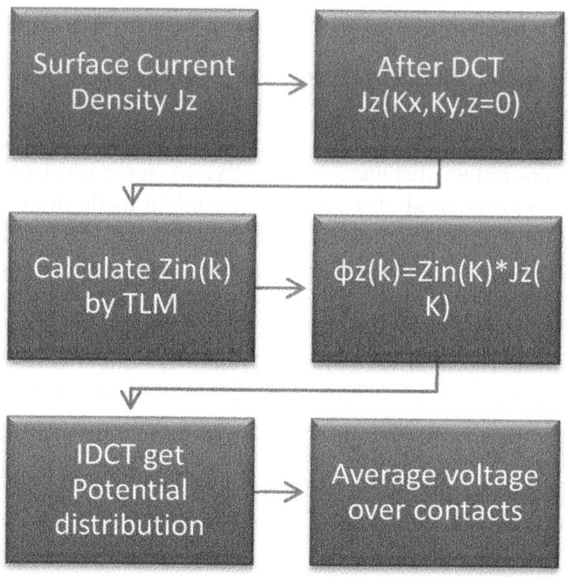

Figure 3.17. *General steps of the algorithm (inverse DCT (IDCT))*

3.2.2. *Model validation*

Impedance parameter Z will be extracted from a multilayer substrate by 3D-IE and finite element method (FEM), respectively.

First, the impedance will be calculated exactly by our algorithm above. For numerical simulation, we use a well-known and robust multiphysics simulator called COMSOL (electrical, thermal, mechanical stresses and their couplings). It is also a dedicated tool for electromagnetic analysis. It essentially uses Galerkin-type algorithms. Typically, a 3D simulation can take a few tens of minutes or even several hours.

In our method, we inject a constant current density at any point of the contact matrix; we use Millman's theorem which is applied throughout the contact. The contact voltage is the average value of discrete contact element voltages, calculated by:

$$V = \frac{\sum_{n=1}^{N} j\omega c \varphi_n}{\sum_{n=1}^{N} j\omega c} = \frac{1}{N} \sum_{n=1}^{N} \varphi_n \qquad [3.9]$$

where φ_n is the substrate voltage at a node and N is the number of points of contact. Thus, the substrate contact plate is considered to be an injection of the constant current density and not a uniform potential.

We are quite aware that the quasi-electromagnetostatic modeling framework cannot work as perfectly up to 10 THz (against experiments), which implies a wavelength of about 10 μm for Si; this wavelength is of the same order of magnitude as inter-contact distances. Currently, we have not explicitly considered nor is it realistic to go beyond 200 GHz (our simulations do not exceed 20 GHz) in this analysis, but we think it is interesting to compare, ab initio, the FEM to test the robustness of our algorithm (fundamental purpose) beyond this frequency.

As an example, a P+ /P region of a 0.35 μm BiCMOS process is chosen and decomposed into 12 layers from bottom to top with thicknesses t1, t2 ... t12 and conductivities σ1, σ2 ... σ12, which are shown in Figures 3.18 and 3.19. The relative permittivity of all layers is 11.9.

The frequency characteristic of impedance will be sought in the range of 10^8–10^{13} Hz and the distance in the X direction between contacts (or TSVs) varies from 1 to 16 μm.

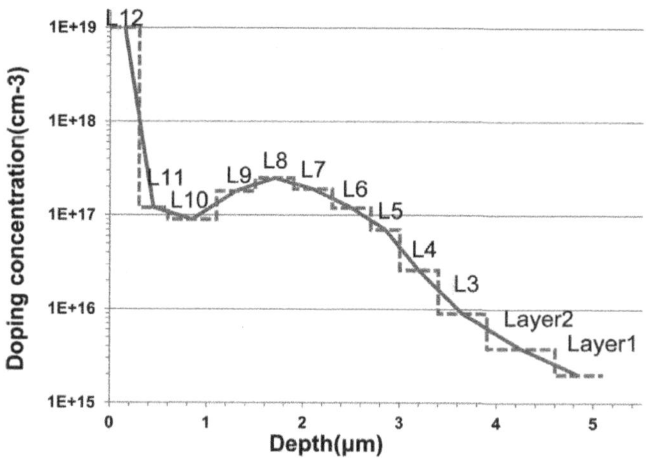

Figure 3.18. *Doping profile: 0.35 µm technology (p+/p region). For a color version of this figure, see www.iste.co.uk/gontrand/analog2.zip*

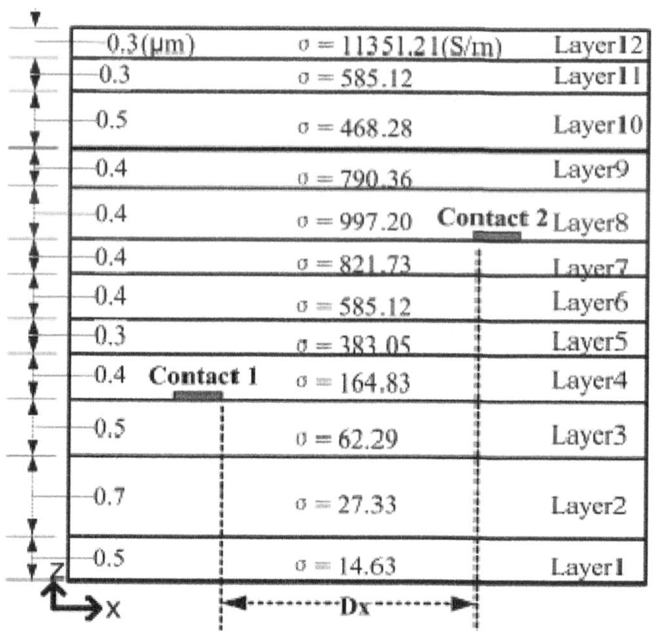

Figure 3.19. *Characteristic of substrate geometry in Z-direction (region) and physical property (conductivity) with two integrated contacts; relative permittivity is always 11.9. Buried contacts in layers 4 and 8. The dimensions in the X and Y direction for the substrate and the contacts are the same as in Figure 3.18(a)*

One contact is embedded in the lower surface of layer 4 and the other is present on the lower surface of layer 8. Like the above test for surface contacts, the lateral separation distance (Dx) of contacts is 6 μm and the frequency is $1e^8$–$1e^{13}$ Hz. Two different parameters of the COMSOL port are applied for analysis: the constant current density (CD Constant), which injects a uniform current density to the contact port, and constant voltage (constant V), which injects a uniform voltage to the contact.

The frequency results of these three methods are shown in Figure 3.20(a). According to the results of this Figure 3.20, there is a good adjustment of the frequency characteristics for the "COMSOL CD Constant" method.

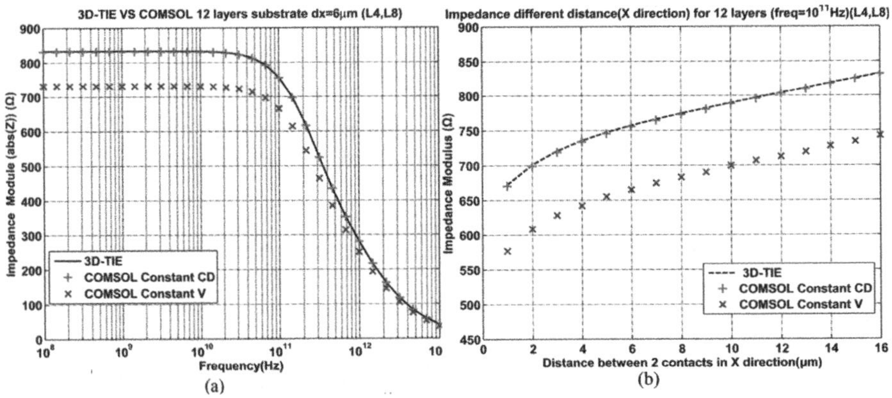

Figure 3.20. *(a) Impedance of 12 substrate layers with separation distance Dx = 6 μm for the two contacts that are embedded in the substrate. One is in layer 4 and the other in layer 8. The results are impedance modulus: (a) Dx = 6; (b) impedance modulus for different distances between contacts. (f = 1011 Hz). For a color version of this figure, see www.iste.co.uk/gontrand/analog2.zip*

Figure 3.20(b) shows impedance results with the variable contact separation distance at the frequency of $1e^{11}$ Hz.

Figure 3.21 shows the potential distribution of the upper surface of the substrate. 3D-IE and COMSOL results are in accordance with each other.

Based on the contact pattern, one of the contacts is extruded into the substrate to obtain a TSV of 5.1 μm that extends from layers 12 to 1; the TSV contact pattern is shown in Figure 3.22(a).

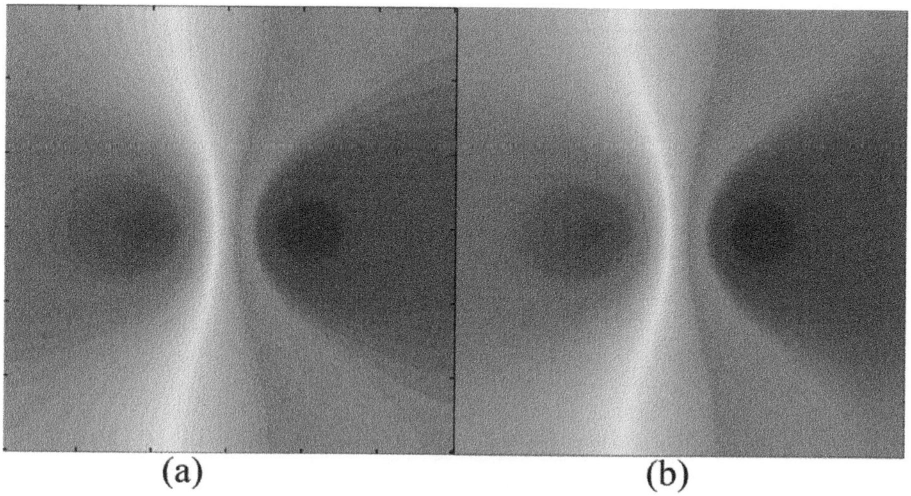

Figure 3.21. *Substrate surface potentials obtained from 3D-IE (a) versus COMSOL (b), for contacts embedded in the 12-layer substrate, in layers 4 and 8. For a color version of this figure, see www.iste.co.uk/gontrand/analog2.zip*

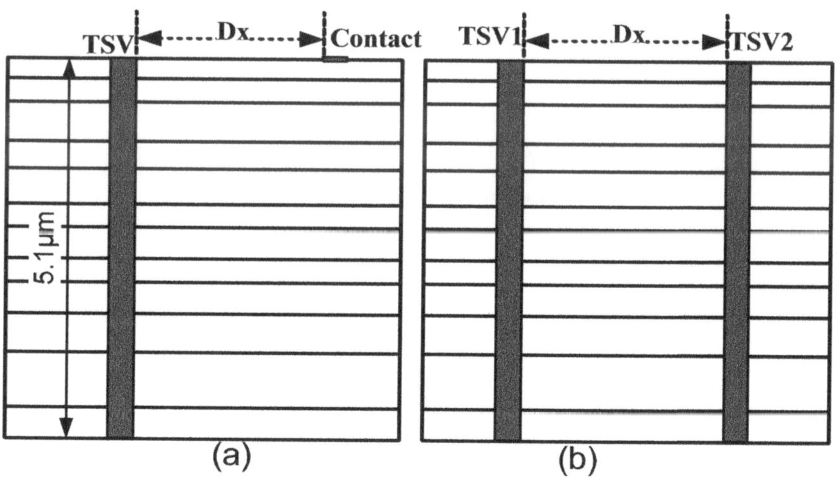

Figure 3.22. *Model schematic: (a) TSV-contact and (b) TSV-model. The TSV dimension is 2 µm × 2 µm × 5.1 µm in X, Y and Z. For a color version of this figure, see www.iste.co.uk/Gontrand/analog2.zip*

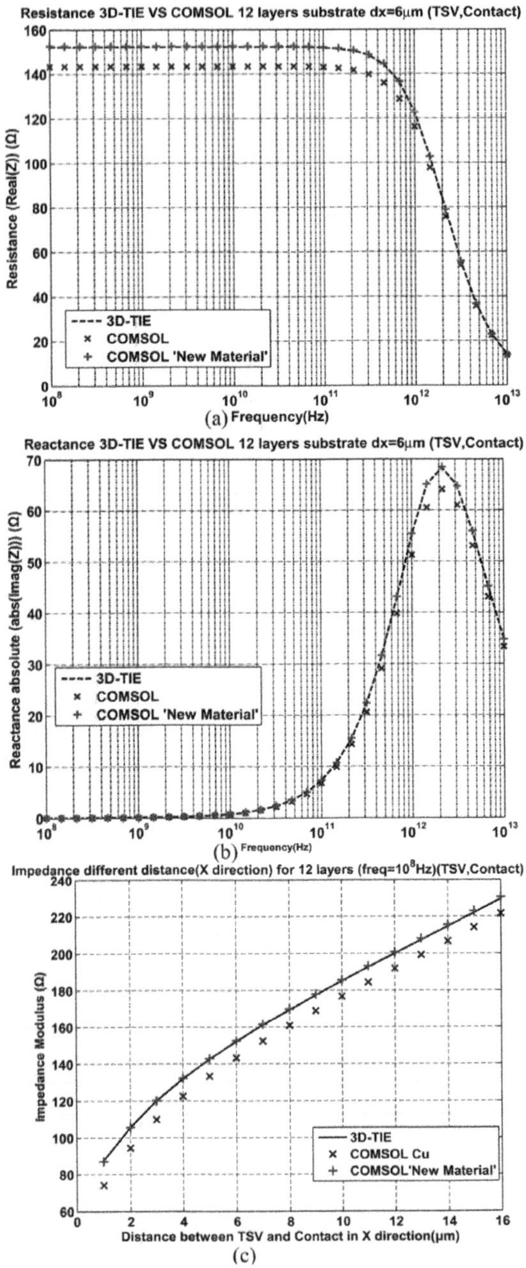

Figure 3.23. *Resistance (a) and absolute value of reactance (b) of 3D-IE and COMSOL between TSV and contact dx = 6 µm under frequencies ranging from 1e^{8} to 1e^{13} Hz. For a color version of this figure, see www.iste.co.uk/gontrand/analog2.zip*

Similarly, the separation distance of TSVs is set at 6 μm and the frequency varies from 10^8 to 10^{13} Hz. The results in Figures 3.23 and 3.25 show a good adjustment of frequency characteristics for the proposed method and COMSOL in terms of resistance and reactance. COMSOL's "New Material" method can in fact consider anisotropic conductivity; this new material of the via has the same conductivity as that of copper in the Z direction, but has the conductivity of that of the substrate layers Si in the X and Y direction (in our analytical method, we currently have lateral conductivities at the via: ε_{SI} (in fact in these x,y directions, there is the capacitive effect of the oxide as well as of a possible deserted zone (which can be understood using the Poisson equation, in cylindrical coordinates, if the via is itself cylindrical) of silicon adjacent to the oxide layer, all the more so since the doping is low). The classic COMSOL method takes an isotropic conductivity, that of copper (by default $\varepsilon_{Cu=}\varepsilon_0$)., for the via. We will need to consider these aspects more closely.

Second, we changed the "contact" distance Dx (Figure 3.25(b)) from 1 to 16 μm, for a frequency equal to $1e^8$ Hz and tested the impedance between TSV and contact. An anisotropically conductive material is also used for comparison. Figure 3.25 shows a strong agreement between our method and the COMSOL simulation. "New material" shows the potential between TSV and contact of COMSOL using the anisotropic material.

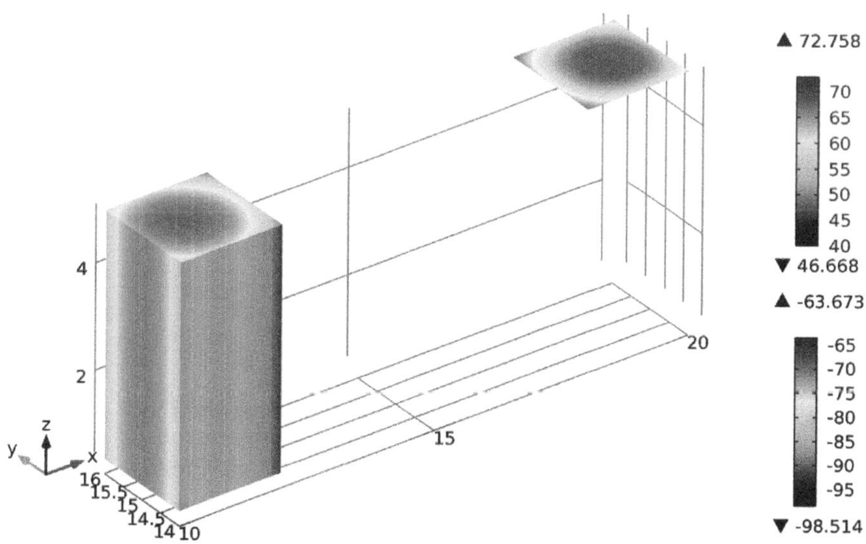

Figure 3.24. *Potential: TSV-contact model with a TSV in an anisotropic material, which has the same conductivity as copper in the Z direction and has the same conductivity components as the silicon substrate in the X and Y directions. For a color version of this figure, see www.iste.co.uk/gontrand/analog2.zip*

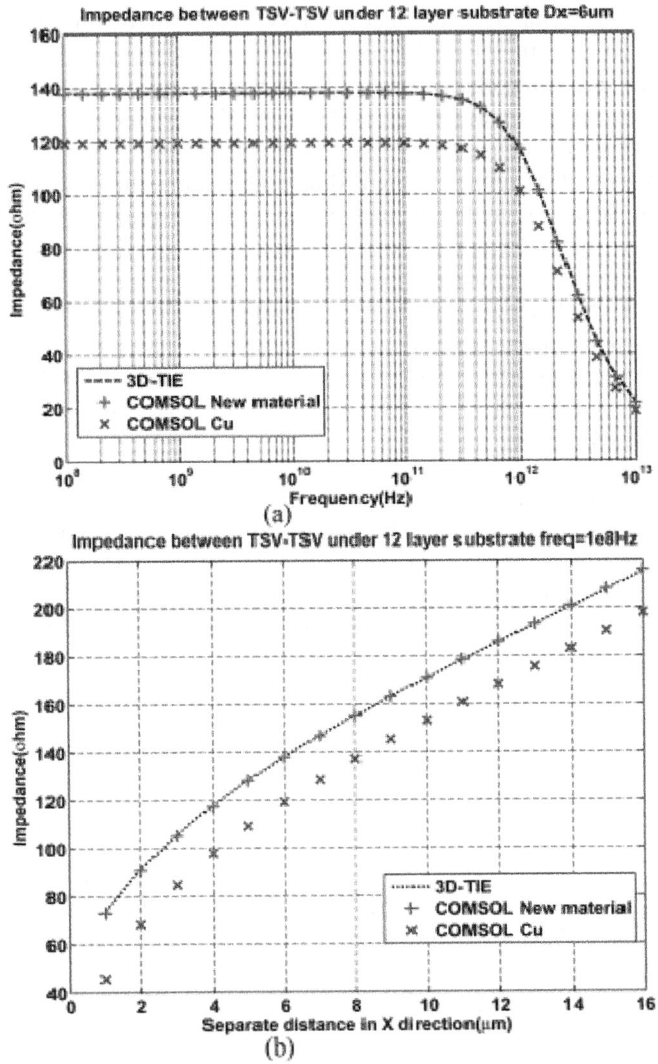

Figure 3.25. *Impedance modulus: 3D-IE and COMSOL between TSVs: (a) a function of frequency with Dx = 6 μm; b) impedance for different separations of 3D-IE versus COMSOL TSVs. For a color version of this figure, see www.iste.co.uk/gontrand/analog2.zip*

Based on the TSV contact model: vertical extrusion of two contacts to obtain two TSVs, which extend from layer 13 to layer 1: a TSV-TSV model can be constructed; its stretched section is illustrated in Figure 3.27.

The separation distance is set at 6 μm, with the frequency range extending from 10^8 to 10^{13} Hz (see Figure 3.31(a)). And we set the frequency to 10^8 Hz, changing the distance between two the TSVs from 1 to 16 μm; we extract the impedance between the two TSVs. Impendences (COMSOL and 3D-IE) resulting from this TSV extraction, for these variable distances, are presented in Figure 3.31(b).

In practice, the TSV has an oxide sleeve, insulation, so a more realistic model is proposed (see Figure 3.26).

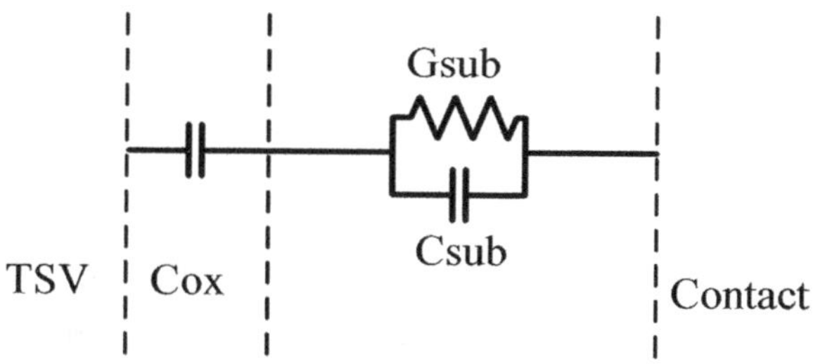

Figure 3.26. *Improvement of the equivalent model of coupling between TSV and contact*

To compare with the old method, a capacitance effect can be added to each TSV piece as shown in Figure 3.26 (eventually, we will add a desertion capacitance in series with this oxide capacitance for low dopings, if necessary, in silicon).

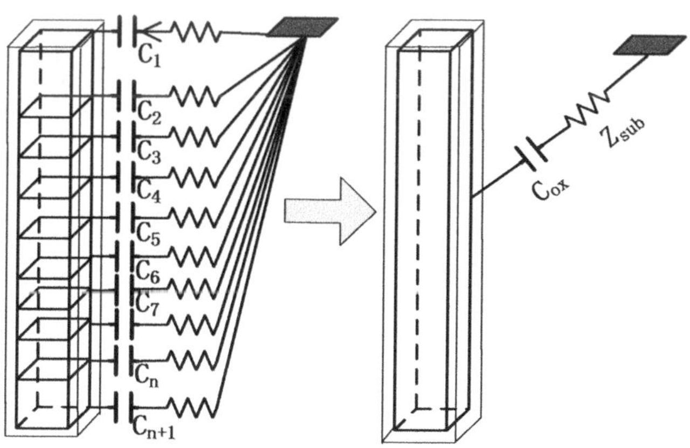

Figure 3.27. *TSV/contact model diagram, wherein the TSV has an oxide coating. C_{ox} is the capacitance of the oxide layer surrounding the TSV; Z_{sub} is the impedance of the substrate*

So, this investigative step is, on the one hand, the calculation of the capacitance induced by the oxide layer; on the other hand, we have to insert it correctly into the model.

Unlike the other method based on Green's kernels, which can only calculate for rectangular contacts, 3D-IE can be used in a more "flexible" way, for example, using disk, crown, rectangle, T-shaped and so on. Figure 3.28 shows some examples submitted to our 3D-IE.

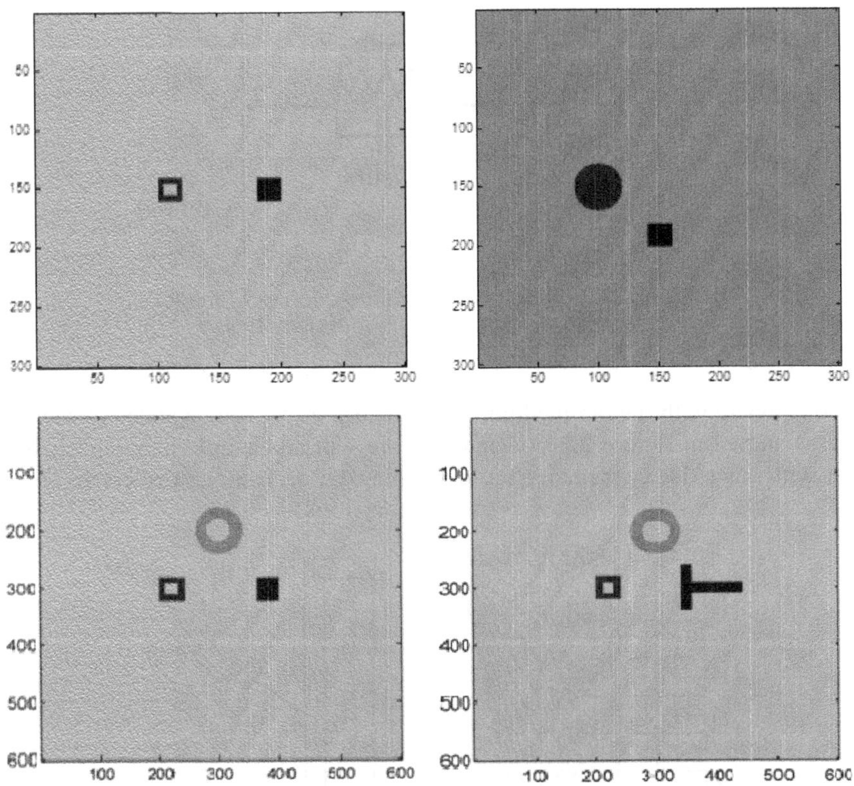

Figure 3.28. *Different contact shapes that can be considered in 3D-IE. For a color version of this figure, see www.iste.co.uk/gontrand/analog2.zip*

If there are trenches and/or wells in a substrate, the layer homogeneity hypothesis becomes invalid, and Green's function cannot be used directly. Extraction of the traditional substrate coupling based on the Green's kernel technique is limited to planar structures and analytical models are generally difficult to develop to account for lateral variations in dielectric constant.

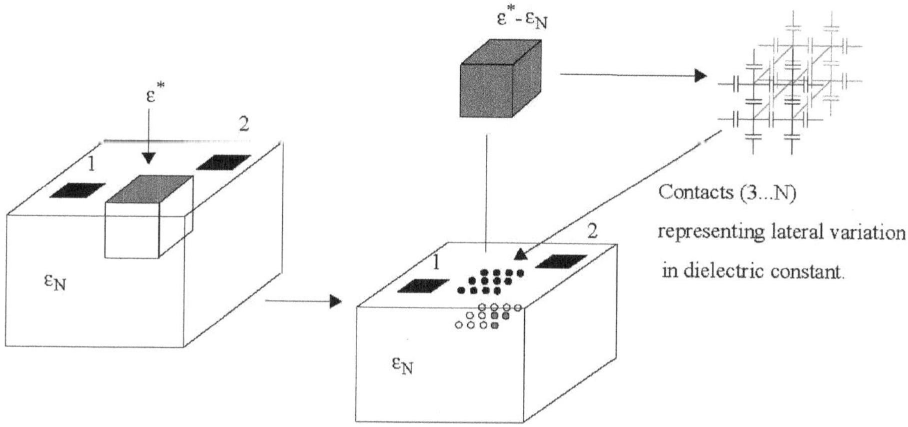

Figure 3.29. *A technique for modeling a lateral variation of dielectric constants (see superposition theorem)*

3.2.3. *Interconnections: compact models*

3D integration is considered to be a promising strong solution. However, due to high-density integration, 3D integration implies that the overall electrical context should be considered as current paths or couplings between chip elements. Thus, in order to accurately evaluate the performance of the 3D system in the design process, electrical compact models are particularly necessary for 3D interconnections.

Compact 3D interconnection models are available, including medium density TSV; they are reliable for low and "medium" frequencies (up to 20 GHz). Parametric 3D electromagnetic simulations and extractions are performed on test structures. This extraction method is experimentally validated in the case of a coplanar waveguide deposited on a high-resistance substrate. RF behaviors of test structures are also studied. Finally, the conclusion summarizes the key points of the chapter and describes future work.

Electrical modeling, in a global context, will be illustrated in this part for coplanar waveguides and TSV chains. While the substrate is highly conductive for coplanar waveguides submitted to us (see CE3-LETI), it can be modeled as a very low resistance, or even a simple node. For each coplanar line, the vertical current path in the epitaxial layer is modeled by a resistor (Repi) in parallel with a capacitor (Cepi). Environmental modeling also contains oxide layer capabilities (Cox) insulating lines from the substrate.

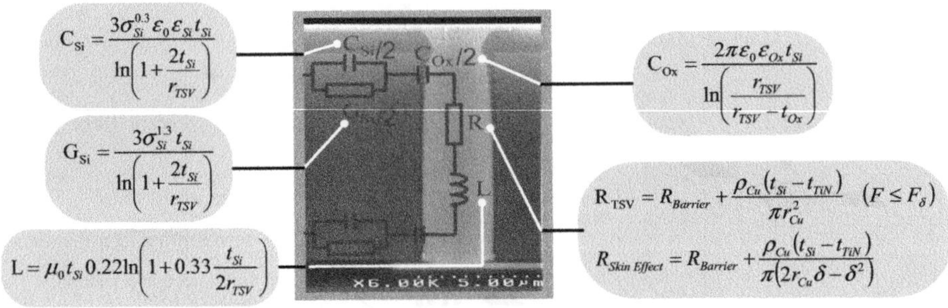

Figure 3.30. *A high-density analytical TSV model. For a color version of this figure, see www.iste.co.uk/gontrand/analog2.zip*

Because system descriptions can become complex when considering the electrical context and distribution of compact models, we have developed a 3D extraction tool under Matlab, 3D transmission line extractor (3D TLE), based on our modeling approach and which integrates the substrate extraction algorithm (3D I.E.). Through a hierarchical syntax and specific declarations, the user describes geometric and technological data in a text file defining the layers, the type of elements included in the system, the connectivity between elements that are "instantiated" components and interactions of elements (coupling, interactions with layers for modeling current flow).

Figure 3.31 shows the 3D TLE tool environment. The extraction tool generates a SPICE type "netlist" from this text file, containing the RLCG system, the electrical description, exportable to CAD tools such as Agilent Technologies ADS. The viability of our 3D TLE was verified by S-parameter comparisons for data systems between their SPICE 3D TLE netlists and their ADS-designed schemas. As expected, results show that answers are "the same" but describing the structure using 3D TLE seems easier and faster than designing using CAD tool. 3D TLE also provides the 3D image of the system.

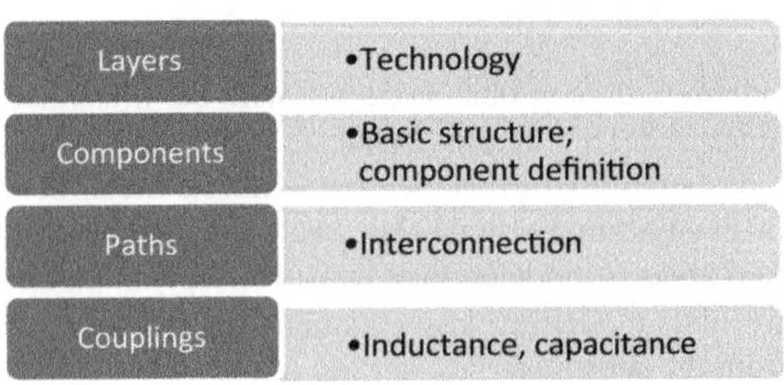

Figure 3.31. *Text file structure to define a specific substrate structure*

3.2.4. *Validation: test structures*

At first glance, a simple structure for coplanar waveguide (CPW) is tested as shown in Figure 3.32.

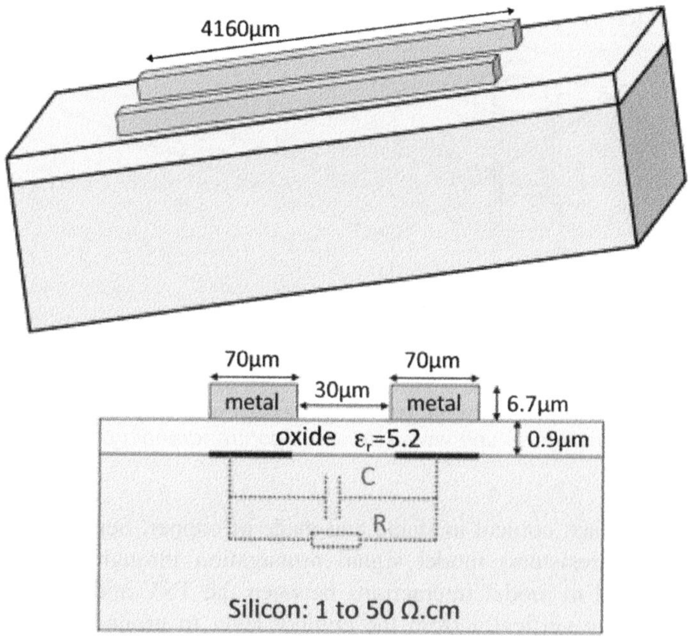

Figure 3.32. *Diagram of a CPW structure used for testing (source: LETI). For a color version of this figure, see www.iste.co.uk/gontrand/analog2.zip*

Modeling is validated in the RF frequency domain through comparisons of S parameters between measurements made on test structures for two different configurations and simulation results of their equivalent electrical models. The test structures used will be presented first, then the comparison between the test and the simulation will be given.

The structure is U-shaped such that there is a signal line at the RDL level connected to 2 TSVs going up to the BEOL level where there are two ground lines. Copper ground lines are located on either side of the BEOL access point of TSVs. The substrate is covered with an epitaxial layer to which a silicon oxide layer is added to insulate copper lines from the substrate. Similarly, an SiO_2 layer separates the signal line from the RDL level of the substrate. The structure is therefore as illustrated in Figure 3.33.

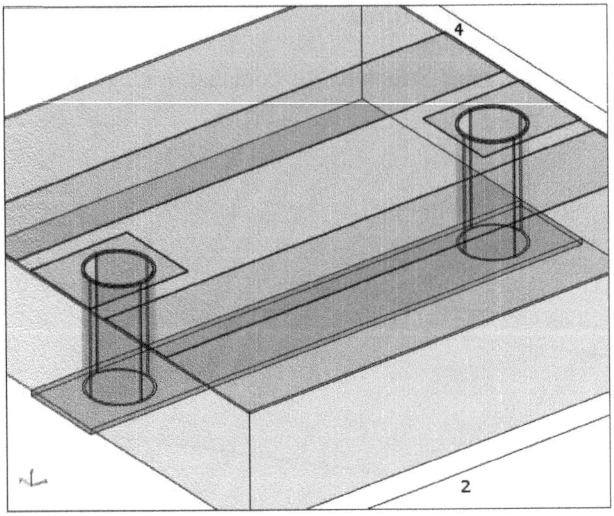

Figure 3.33. *"U" structure: a signal line located at the RDL level + 2 TSV + 2 ground lines located at the BEOL level. For a color version of this figure, see www.iste.co.uk/gontrand/analog2.zip*

TSVs are, in fact, conical in shape and made of copper. Serial elements (partial inductances and resistors) model signal propagation through the TSV. Parallel elements are used to model interactions between the TSV and their environment. Currents propagate vertically from the epitaxy layer to ground lines. This path is modeled by a resistor in parallel with a capacitor, to which the capacitance of the thin oxide layer separating TSV and substrate and that relating to the SiO_2 layer separating ground lines of the substrate are added in series. The presence of a ground line on either side of TSVs implies modeling two paths.

Ultimately, this amounts to the TSV considering a four-port structure. Two are used to connect RDL and BEOL levels, while the other two are used to reflect the interaction between the TSV and the two ground lines of the BEOL level. Note that TSV, modeled under ADS here, is a subcomponent that can be included in other models.

Test structures are made using a highly conductive silicon substrate having a thickness of 120 μm, a resistance ρ = mΩ•cm and a relative permittivity $\varepsilon_r = 11.7$. A thin epitaxial layer (thickness: 4.5 μm; resistance: 10 Ω •cm; permittivity: 11.7) is deposited on the substrate. Lines at the BEOL level, for TSPE TSV and chains, are insulated from the substrate by an oxide layer (thickness: 0.8 μm; relative permittivity: 6.2). For TSV chains, an oxide layer (thickness: 1 μm; permittivity: 6.5) is added to the back of the structure to insulate the BRDL line from the substrate. All layers used for PTC and TSV chain structures are shown in Figure 3.36 with their respective characteristics. The BEOL level line sections have a thickness of 0.8 μm and 7 μm for the RDL level.

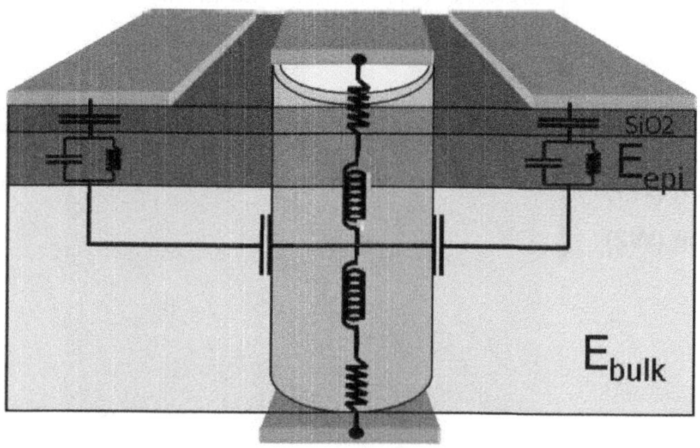

Figure 3.34. *TSV RLCG model. TSV + current paths between the TSV and the two ground lines of the BEOL level located on either side of the TSV. For a color version of this figure, see www.iste.co.uk/gontrand/analog2.zip*

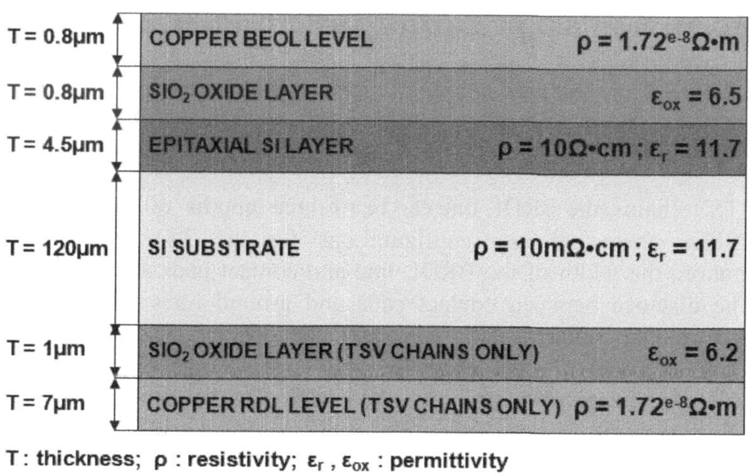

Figure 3.35. *Geometric and technological data of layers in the CPW and TSV chain structures. For a color version of this figure, see www.iste.co.uk/gontrand/analog2.zip*

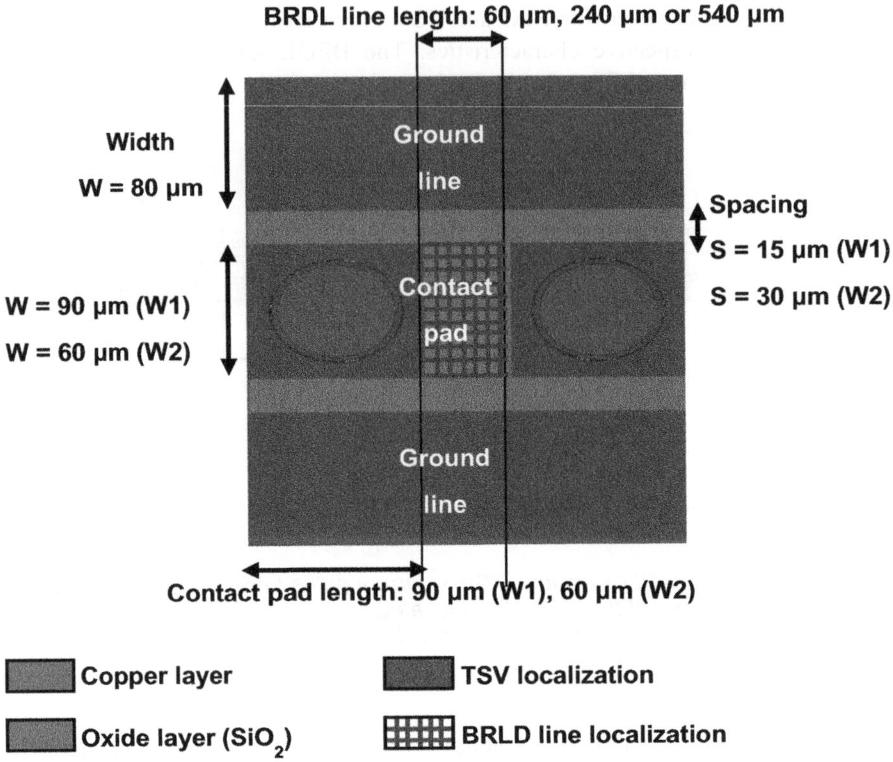

Figure 3.36. *Top view of a 2xTSV chain structure configuration. For a color version of this figure, see www.iste.co.uk/gontrand/analog2.zip*

For TSV chains, the BRDL line can be of three lengths: 60, 240 and 540 µm. As with TSPEs, there are two configurations for the TSV chain. In the W1 configuration, the width of the BRDL line and contact pads are 90 µm (60 µm for W2). The distance between contact pads and ground lines also depends on the configuration. Gap values are the same as CPW structures, as are baseline widths. Figure 3.43 shows the top view of the 2xTSV chain, and Figure 3.44 shows the sectional view. For the sake of clarity, the BRDL line and contact pad locations are indicated.

Different colors represent different materials as presented in Table 3.1.

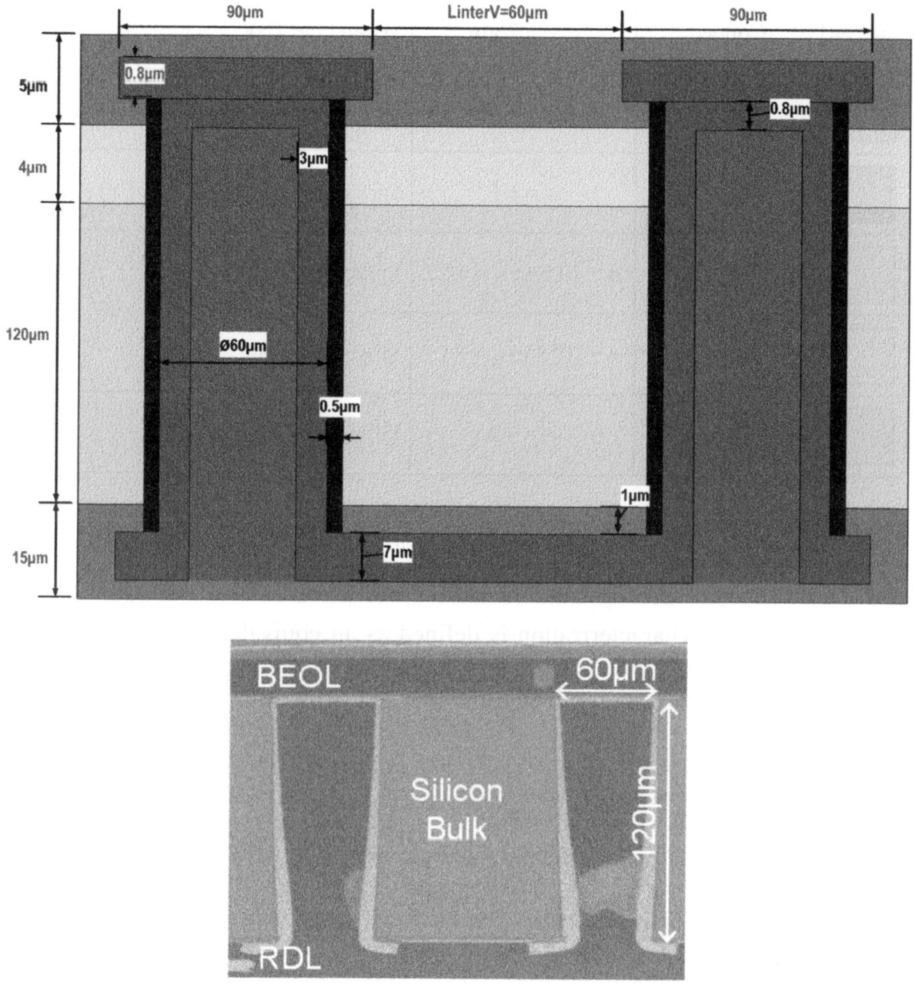

Figure 3.37. *SEM cross-section of the average TSV density used in TSV chains (configuration W1). For a color version of this figure, see www.iste.co.uk/gontrand/analog2.zip*

Color	Material	Physics property	Description
	Copper	$\sigma = 5e7S/m$	M1 and the TSV
	SiO_2	$\varepsilon_r = 6.6$	Surface oxide layer and back oxide layer
	SiO_2	$\varepsilon_r = 6.5$	Surface area of the TSV, called a wall
	Epitaxial silicon	$\sigma = 10$ S/m $\varepsilon_r = 11.7$	Epitaxial layer
	Silicon	$\sigma = 10^4$ S/m $\varepsilon_r = 11.7$	Silicon substrate
	SINR	$\varepsilon_r = 3.4$	Filling the TSV

Table 3.1. *Electrical parameters for various materials involved. For a color version of this table, see www.iste.co.uk/gontrand/analog2.zip*

Interconnection characterization is defined as an equivalent electrical model to be extracted from measurements. For example, the characteristic impedance and a propagation exponent or distributed elements R, L, C and G are extracted as a function of frequency from the equivalent model of the telegraph equation and from measuring the diffusion matrix, containing S parameters (S: scattering), of this interconnection using a vector network analyzer.

In practice, this interconnection, which can be called a device under test (DUT), is generally in a complex test environment; it does not allow for extracting its parameters by a single measurement. In fact, this interconnection is buried in a cell to connect to the measurement system (generally, we find contact pads, access sections, cables, connection lines, etc.). "Deembedding" methods (stripping, peeling) are used to avoid environment influence and obtain intrinsic characteristics of the DUT, which is the interconnection for us.

Therefore, the modeling approach is then based on analysis of test structures by means of a 3D electromagnetic tool using the FEM to understand physical and electrical phenomena that occur in 3D, as for interconnections, during signal propagation, and finally on the use of a transmission line method.

Figure 3.38. *Measurements with a vector network analyzer (VNA). For a color version of this figure, see www.iste.co.uk/gontrand/analog2.zip*

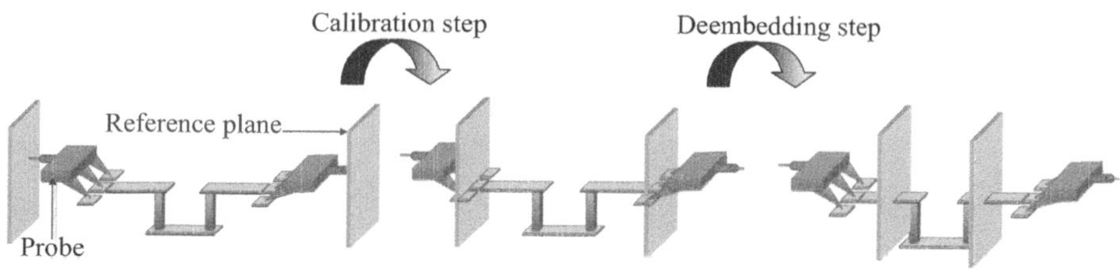

Figure 3.39. *Effects of calibration and removal of coating. For a color version of this figure, see www.iste.co.uk/gontrand/analog2.zip*

3.2.5. *Numerical simulations*

COMSOL simulation is carried out via the "radio frequency" module. A major difference between quasi-static and very high frequency modeling is that formulations depend on the "electrical size" of a structure. This dimensionless measurement is the ratio of the greatest distances between two points on the structure divided by the nominal wavelength of electromagnetic fields. Quasi-static formulations are appropriate for simulations of structures with an electrical dimension in the range up to 1/10. The physical assumption of these situations is that wave propagation delays are small enough to be overlooked. Thus, phase shifts or phase gradients in fields are caused by materials and/or by inductive or capacitive conductor couplings, rather than being caused by propagation delays. For electromagnetism, electrostatics, magnetostatics, in quasi-statics, we use the AC/DC module, a multi-physics module of COMSOL. When propagation delays become significant, it is necessary to use complete Maxwell equations for high-frequency electromagnetic waves: the "full wave" module (which we do not yet have). They are suitable for structures of electrical size 1/100 and above. Thus, an overlap area exists where we can use both quasi-static and "Maxwell" models (rotE, rotH).

We set the outer surface boundary with the "scattering boundary conditions".

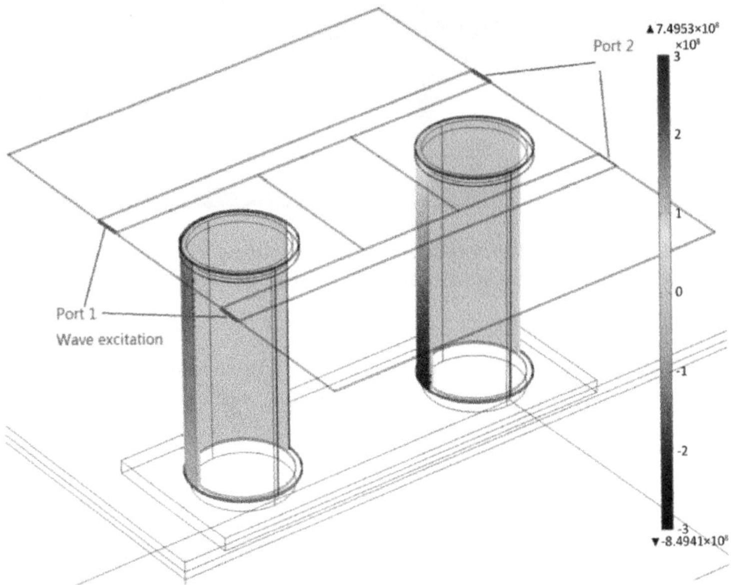

Figure 3.40. *Lumped ports. For a color version of this figure, see www.iste.co.uk/gontrand/analog2.zip*

Lumped ports are fixed as shown in Figure 3.40. The characteristic impedance is 50 Ω for each port.

Mesh sizes and shape play a very important role in finite element simulation. Optimal mesh leads to accurate results with less time and memory consumption. A manual mesh can also be used in addition to a physical control mesh. A viable mesh could be created by swept mesh technology and layered mesh technology. The mesh is shown in Figure 3.41.

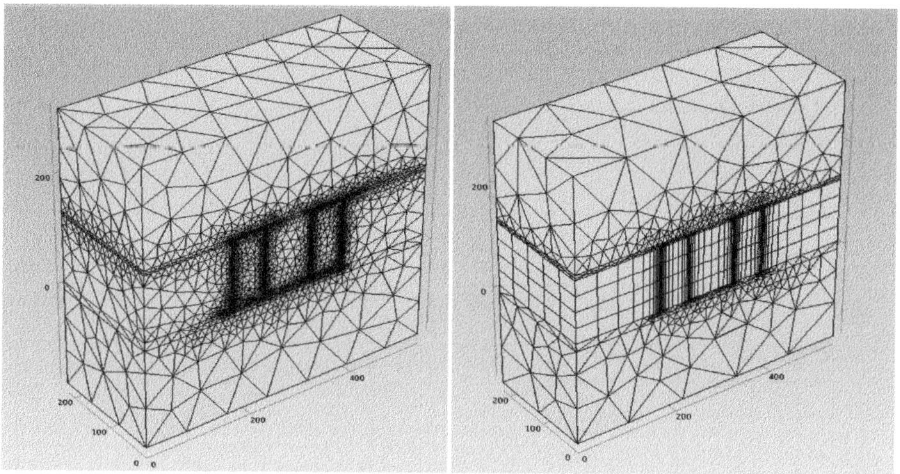

Figure 3.41. *Physical mesh control (left) and manual use of Swept Mesh Tech mesh (right)*

With an interest in the "skin effect" on the outer surface of the TSV, a suitable "Layered Mesh" could be used on the surface of the TSV to improve the accuracy of the result. This layered mesh could produce a very fine mesh on the "near surface" domain.

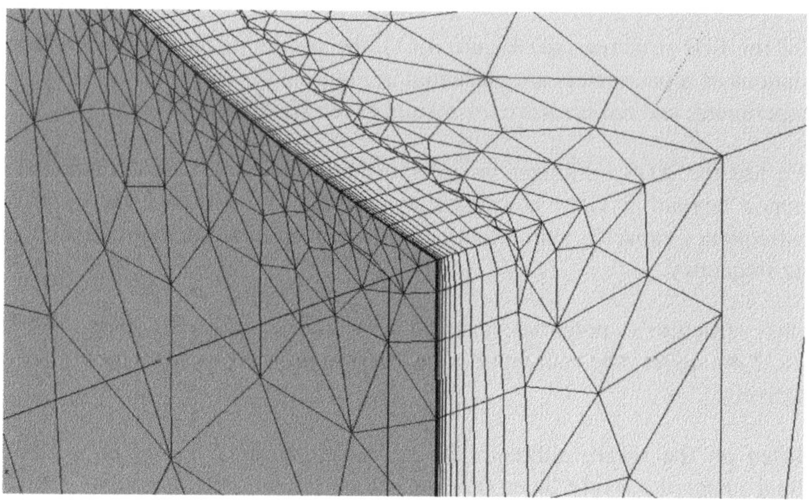

Figure 3.42. *Mesh layers on TSV surface*

3.2.5.1. *Compact test and model results*

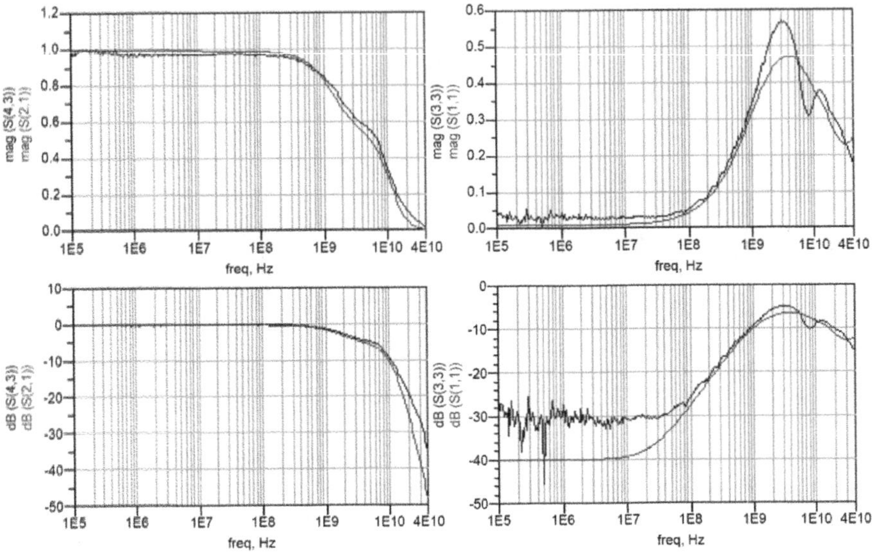

Figure 3.43. *S parameter for the simple CPW structure (S_{21} left; S_{11} right). For a color version of this figure, see www.iste.co.uk/gontrand/analog2.zip*

For the first structure (see Figure 3.41), the results of the experiment and ADS simulations of S parameters are presented in Figure 3.42. Blue lines are the results of the experiment, and red lines are the results of the IE/ADS simulation.

We use the ADS optimizer in order to have the "best" set of parameters; the difference beyond 5 GHz is caused by parameters and depends on frequency (high-frequency capacity, skin effect). Then, we need to refine our model beyond the 5 GHz frequency.

This structure's potential distribution is also studied under COMSOL. Figure 3.44 gives the potential distribution on and below the oxide layer, respectively.

Based on the result, although the potential on the oxide layer is uniform, a potential under the oxide layer (i.e. on top of the substrate) may be found to be nonuniform.

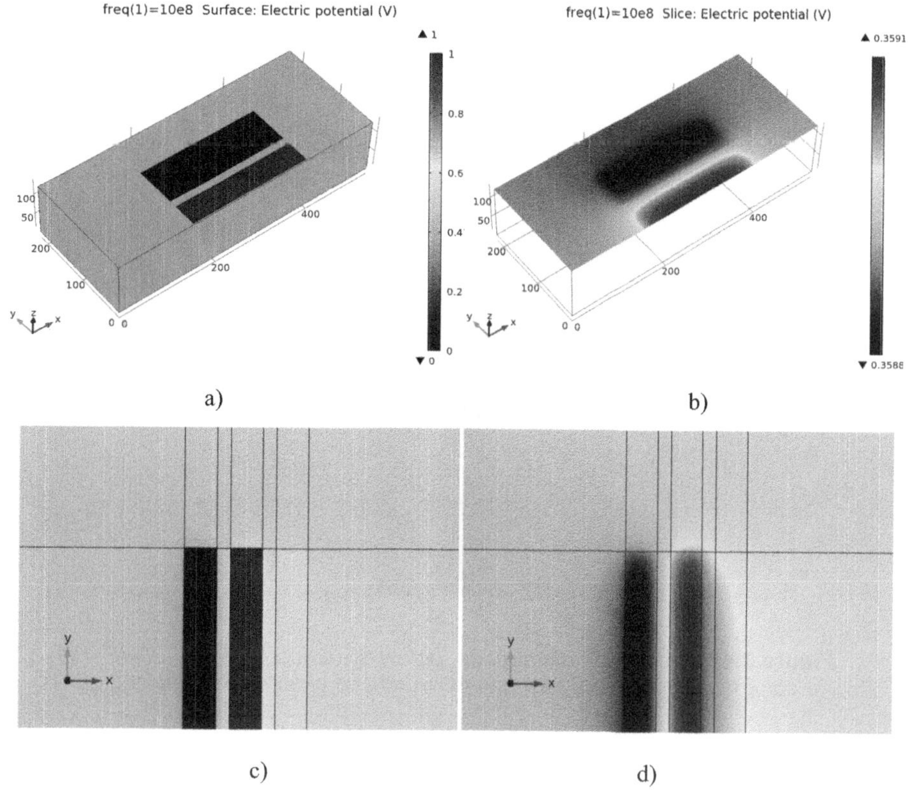

Figure 3.44. *Potential distribution on oxide layer (a and c) and below oxide layer (b and d) (of COMSOL). For a color version of this figure, see www.iste.co.uk/gontrand/analog2.zip*

This is why we did not use a constant voltage on the substrate impedance calculation. A potential distribution below the oxide layer, calculated via 3D-IE for the same structure, is shown in Figure 3.45.

We have to note that the use of the constant current method is more accurate than the constant voltage method on top of the substrate.

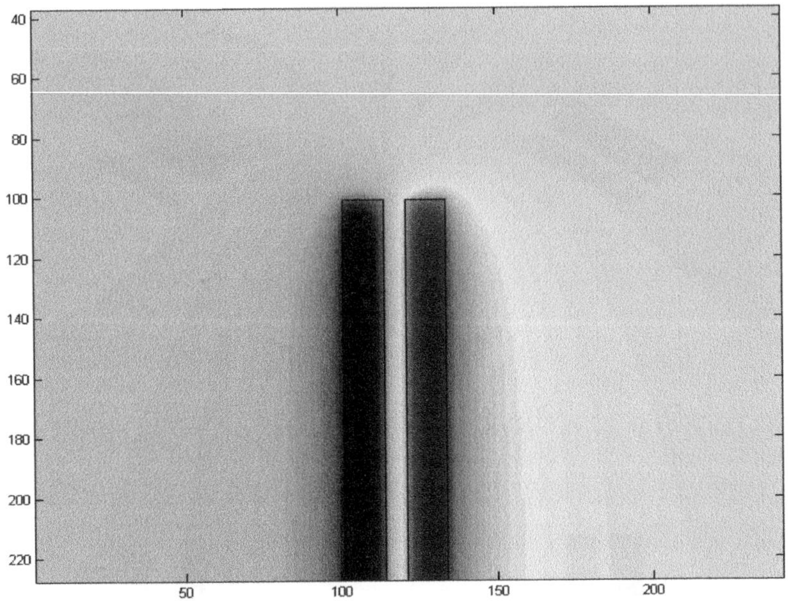

Figure 3.45. *A potential distribution, below the oxide layer, via 3D-IE. For a color version of this figure, see www.iste.co.uk/gontrand/analog2.zip*

Figure 3.46 shows comparison results. RF responses of test structures are plotted with continuous red curves – those of electrical models – and with dotted blue curves (network measurements). Equivalent compact RLCG models proposed for coplanar waveguides and TSV channels show great accuracy for frequencies up to 20 and 10 GHz, respectively. Some peaks can be seen in measurements, but not in "low" frequency simulations (below 1 GHz). These resonances seem to be due to coupling between the substrate of the test structure and its measurement environment.

W2 configuration coplanar waveguides result in better transmission resulting from insertion losses due to higher impedance along the escape path through the substrate. The signal line width is smaller in this configuration, resulting in lower oxide capacity. Nevertheless, insertion losses are dominated by BEOL lines: coplanar resistive losses that increase for longer line lengths.

At very high frequency, the estimation error between RF responses of test structures and their equivalent electrical models is greater than 10% due to neglected substrate coupling effects. The substrate extraction method, combined with the modeling approach, is validated here for a coplanar waveguide at the top of a highly resistive substrate. S parameters proposed by the electrical model were compared, for a frequency range of up to 20 GHz, with responses obtained, for the structure produced, under an FEM 3D electromagnetic simulator. The results of this comparison are shown in Figure 3.54. The error percentage in S_{12} evaluation is relatively high, while losses are in fact practically negligible, as can be seen in the figure.

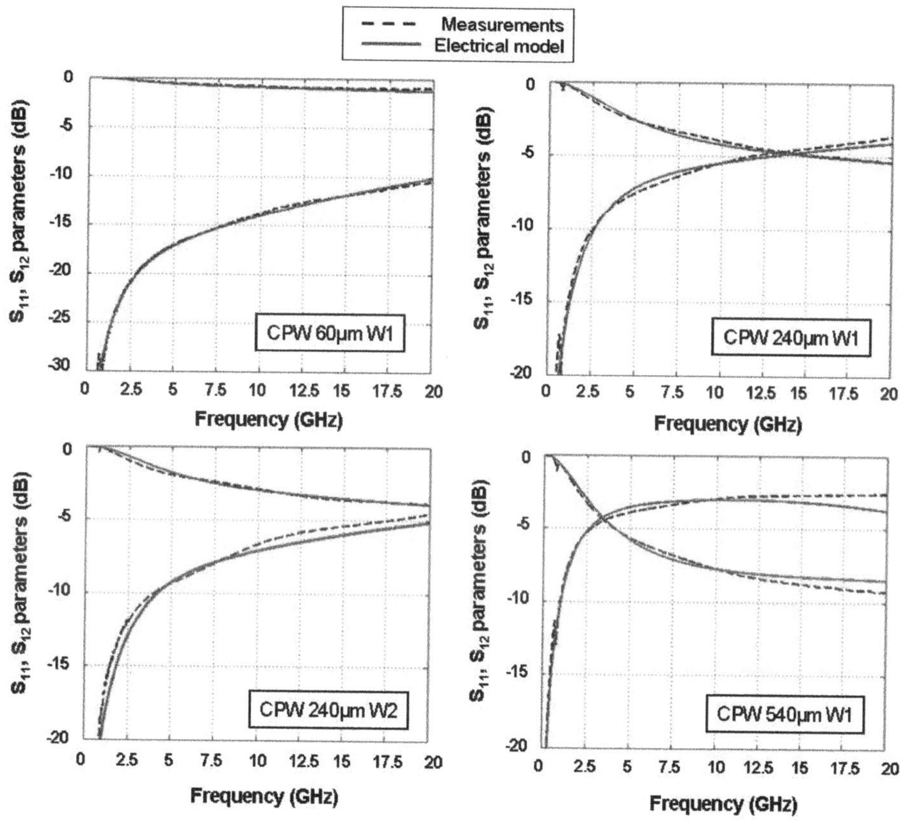

Figure 3.46. *Comparison of S parameters, up to 20 GHz, between measurements made on coplanar waveguide test structures and simulations of their electrical equivalent models for both W1 and W2 and configurations of different coplanar line lengths. For a color version of this figure, see www.iste.co.uk/gontrand/analog2.zip*

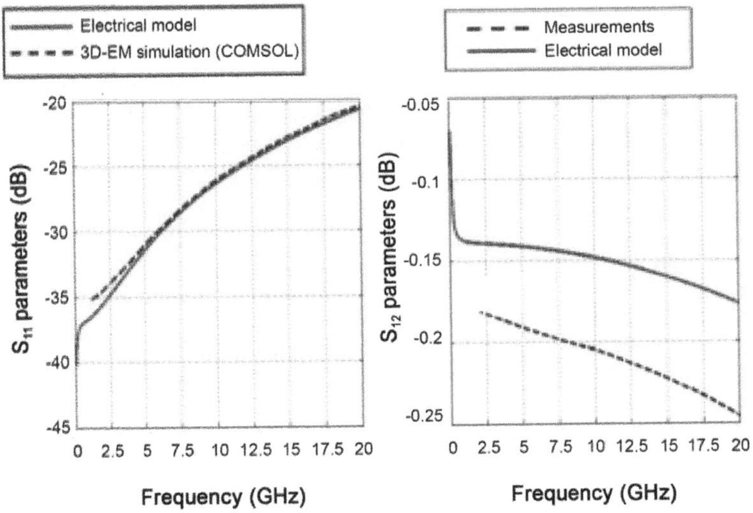

Figure 3.47. *Comparison of S parameters for a 240 m coplanar waveguide system at the top of a highly resistive substrate. The substrate is modeled with DC electrical parameters. Left: S_{11} parameters. Right: S_{12} parameters. For a color version of this figure, see www.iste.co.uk/gontrand/analog2.zip*

3.2.6. *Prospects and future directions*

Here, we shall disccuss future prospects and directions. The study of the integrity of the signal in TSV matrices will be introduced first, then, we shall look at the (electro-)thermal analysis of the TSV structure. The skin effect and eddy currents at the TSV should also be analyzed. A starting point concerning the results on the possible correlation of signal fluctuations, at two distinct points in volume, is presented; this should initiate the study of noise source correlations in 3D circuits, in particular at the "nanoscale" level.

Temporal analyses are carried out: eye diagrams, applied to different TSV matrix configurations, according to the types of signals sent to the TSV near the extremities, to study signal integrity problems in such 3D structures. Results show that the TLE 3D Extractor can be easily integrated into a common EDA environment by providing efficient compact models, for example, to analyze TSV beam performance for high-speed data communication.

In 3D, heat dissipation devices have, a priori, a higher heat density than in a comparable 2D chip in terms of functionality. Figure 3.48. shows, for example, a temperature map in an SRAM: Static_Random_Access_Memory (FEM calculations).

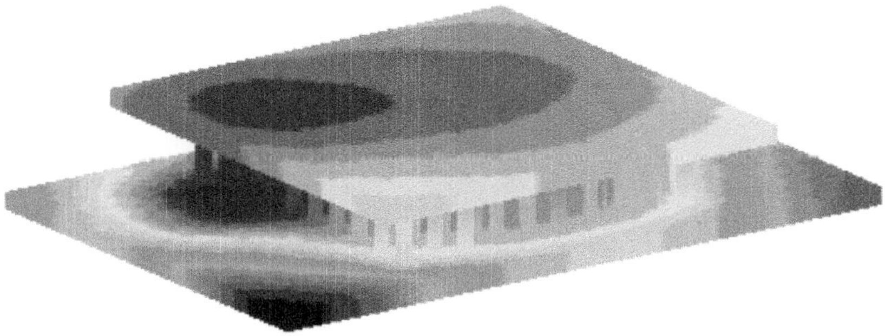

Figure 3.48. *Temperature map in an SRAM. For a color version of this figure, see www.iste.co.uk/gontrand/analog2.zip*

3.3. Conclusion

This volume, dedicated to analog microelectronic circuitry, follows on from Volume 1, which has focused on related devices upstream.

Far from being exhaustive, we offer a doorway to this exciting world. Future entrants must make a sustained effort in this vast field, even if it goes against commonly accepted lessons.

Once the philosophy or even the basics have been acquired, we will be able to get into sums, essentially American, or even visit major university sites, essentially across the Atlantic.

Nec plus ultra: do not hesitate to read concise, clear and educational "applications notes" of famous companies in this field. The "sharpest" readers need to take an interest in AI, which will revolutionize this field (and many others).

References

Bilardi, G., Padovani, R., Pierobon, G. (1983). Spectral analysis of Markov chains functions with applications. *IEEE Transactions on Communications*, 31(7), 853–861.

Cadence (n.d.). Circuit Simulation [Online]. Available at: https://www.cadence.com/en_US/home/tools/custom-ic-analog-rf-design/circuit-simulation.html.

COMSOL (n.d.). COMSOL Multiphysics® Software [Online]. Available at: https://www.comsol.com/.

Gontrand, C. (2014). *Towards a Modelling Synthesis of Two or Three-Dimensional Circuits Through Substrate Coupling and Interconnections: Noises and Parasites*. Bentham Science, Sharjah.

Gontrand, C. (2018). *Micro-nanoelectronics Devices: Modeling of Diffusion and Operation Processes*. ISTE Press, London, and Elsevier, Oxford.

Gontrand, C. (2019). *Power, Thermal, Noise, and Signal Integrity Issues on Substrate/Interconnects Entanglement*. CRC Press, Boca Raton.

Grey, P.R. and Meyer, R.G. (1993). *Analysis and Design of Analog Integrated Circuits*. Wiley, New York.

Hajimiri, A. (1999). *The Design of Low Noise Oscillators*. Kluwer Academic Publishers, Dordrecht.

Millman, J. and Grabel, A. (1969). *Microèlectronique*. McGraw Hill, New York.

Nagel, L.W. (1975). SPICE2: A computer program to simulate semiconductor circuits. PhD Thesis, University of California, Berkeley.

Nagel, L.W. and Pederson, D.O. (1973). SPICE (Simulation Program with Integrated Circuit Emphasis). Report, ERL-M382, University of California, Berkeley [Online]. Available at: http://www.eecs.berkeley.edu/Pubs/TechRpts/1973/22871.html.

Niknejad, A., Gharpurey, R., Meyer, R.G. (1995). Modeling and analysis of substrate coupling in integrated circuits. *IEEE Conference on Integrated Circuits*.

SILVACO (n.d.). Semiconductor Process and Device Simulation [Online]. Available at: https://silvaco.com/tcad.

Sun, F. (2013). Analysis and characterization of substrate and connection couplings in 3D circuits: Towards compact models. PhD Thesis. INSA Lyon, Lyon.

Vapaille, A. and Castagné, R. (1990). *Dispositifs et circuits intégrés semiconducteurs*. Dunod, Malakoff.

Index

T, W

Other titles from

iSTE

in

Electronics Engineering

2023

COURRIER Thierry, QUARTIER Laurent
Ondes Martenot with Tubes

GONTRAND Christian
Analog Devices and Circuits 1: Analog Devices

NGUYEN Thien-Phap
Defects in Organic Semiconductors and Devices

2022

METHAPETTYPARAMBU PURUSHOTHAMA Jayakrishnan, PERRET Etienne, VENA Arnaud
Non-Volatile CBRAM/MIM Switching Technology for Electronically Reconfigurable Passive Microwave Devices: Theory and Methods for Application in Rewritable Chipless RFID

2021

NGUYEN Thien-Phap
Organic Electronics 1: Materials and Physical Processes
Organic Electronics 2: Applications and Marketing

2020

CAPPY Alain
Neuro-inspired Information Processing

FOUQUET François
Noise in Radio-Frequency Electronics and its Measurement

MAHDOUM Ali
CAD of Circuits and Integrated Systems

2018

MURET Pierre
Fundamentals of Electronics 2: Continuous-time Signals and Systems
Fundamentals of Electronics 3: Discrete-time Signals and Systems, and Quantized Level Systems

2017

BUCCI Davide
Analog Electronics for Measuring Systems

MURET Pierre
Fundamentals of Electronics 1: Electronic Components and Elementary Functions

2016

BAUDRAND Henri, TITAOUINE Mohammed, RAVEU Nathalie
The Wave Concept in Electromagnetism and Circuits: Theory and Applications

FANET Hervé
Ultra Low Power Electronics and Adiabatic Solutions

NDJOUNTCHE Tertulien
Digital Electronics 1: Combinational Logic Circuits
Digital Electronics 2: Sequential and Arithmetic Logic Circuits
Digital Electronics 3: Finite-state Machines

2015

DURAFFOURG Laurent, ARCAMONE Julien
Nanoelectromechanical Systems

2014

APPRIOU Alain
Uncertainty Theories and Multisensor Data Fusion

CONSONNI Vincent, FEUILLET Guy
Wide Band Gap Semiconductor Nanowires 1: Low-Dimensionality Effects and Growth
Wide Band Gap Semiconductor Nanowires 2: Heterostructures and Optoelectronic Devices

GAUTIER Jean-Luc
Design of Microwave Active Devices

LACAZE Pierre Camille, LACROIX Jean-Christophe
Non-volatile Memories

TEMPLIER François
OLED Microdisplays: Technology and Applications

THOMAS Jean-Hugh, YAAKOUBI Nourdin
New Sensors and Processing Chain

2013

COSTA François, GAUTIER Cyrille, LABOURE Eric, REVOL Bertrand
Electromagnetic Compatibility in Power Electronics

KORDON Fabrice, HUGUES Jérôme, CANALS Agusti, DOHET Alain
Embedded Systems: Analysis and Modeling with SysML, UML and AADL

LE TIEC Yannick
Chemistry in Microelectronics

2012

BECHERRAWY Tamer
Electromagnetism: Maxwell Equations, Wave Propagation and Emission

BELLEVILLE Marc, CONDEMINE Cyril
Energy Autonomous Micro and Nano Systems

CLAVERIE Alain
Transmission Electron Microscopy in Micro-nanoelectronics

LALAUZE René
Chemical Sensors and Biosensors

LE MENN Marc
Instrumentation and Metrology in Oceanography

SAGUET Pierre
Numerical Analysis in Electromagnetics: The TLM Method

2011

ALGANI Catherine, RUMELHARD Christian, BILLABERT Anne-Laure
Microwaves Photonic Links: Components and Circuits

BAUDRANT Annie
Silicon Technologies: Ion Implantation and Thermal Treatment

BESNIER Philippe, DÉMOULIN Bernard
Electromagnetic Reverberation Chambers

DEFAY Emmanuel
Integration of Ferroelectric and Piezoelectric Thin Films: Concepts and Applications for Microsystems

DEFAY Emmanuel
Ferroelectric Dielectrics Integrated on Silicon

LANDIS Stefan
Nano-lithography

LE CHEVALLIER François, LESSELIER Dominique, STARAJ Robert
Non-standard Antennas

MONMASSON Eric
Power Electronic Converters: PWM Strategies and Current Control Techniques

RAZIK Hubert
Handbook of Asynchronous Machine with Variable Speed

2010

JUFER Marcel
Electric Drives: Design Methodology

LANDIS Stefan
Lithography

PIETTE Bernard
VHF / UHF Filters and Multicouplers

2009

DECOSTER Didier, HARARI Joseph
Optoelectronic Sensors

FABRY Pierre, FOULETIER Jacques
Chemical and Biological Microsensors / Applications in Fluid Media

GAUTIER Jacques
Physics and Operation of Silicon Devices in Integrated Circuits

MOLITON André
Solid-State Physics for Electronics

PERRET Robert
Power Electronics Semiconductor Devices

SAGUET Pierre
Passive RF Integrated Circuits

DE SALVO Barbara
Silicon Non-volatile Memories / Paths of Innovation

2008

CHARRUAU Stéphane
Electromagnetism and Interconnections

MEUNIER Gérard
The Finite Element Method for Electromagnetic Modeling

2007

PLACKO Dominique
Fundamentals of Instrumentation and Measurement

RIPKA Pavel, TIPEK Alois
Modern Sensors Handbook

Printed and bound by CPI Group (UK) Ltd, Croydon, CR0 4YY
13/09/2026
14961042-0004